Essential Soil Science

A clear and concise introduction to soil science

M. R. Ashman and G. Puri

Blackwell
Publishing

BLACKWELL PUBLISHING
350 Main Street, Malden, MA 02148-5020, USA
108 Cowley Road, Oxford OX4 1JF, UK
550 Swanston Street, Carlton, Victoria 3053, Australia

First published 2002 by Blackwell Science Ltd
9 2010

Library of Congress Cataloging-in-Publication Data

Ashman, M. R. (Mark R.)
  Essential soil science: a clear and concise introduction to soil science/
  M. R. Ashman and G. Puri.
    p. cm
  Includes bibliographical references (p.) and index.
  ISBN 978-0-632-04885-4 (pb.: alk. paper)
    1. Soil Science. I. Puri, G. (Geeta). II. Title.

S591.A84 2001
631.4—dc21

2001043210

A catalogue record for this title is available from the British Library.

Set in 10/13 pt Palatino
by Graphicraft Ltd, Hong Kong
Printed and bound in Singapore
by Ho Printing Singapore Pte Ltd

For further information on
Blackwell Publishing, visit our website:
www.blackwellpublishing.com

Contents

Preface

This book has been written for students who need to understand quickly the basic principles of soil science. Studying a new subject can be a hard and tedious business. Things are not made any easier by the fact that these days very few students have the luxury of studying a single subject, such as 'soil science', full-time. The growth of modular degree schemes has meant that successful study has become something of a balancing act, often between completely different subject areas. This is especially true of soil science, which is taught over a range of courses, including engineering and ecology. This book was written with this problem in mind. Every attempt has been made to assume no previous scientific knowledge, to ensure new concepts are explained simply, often using examples of everyday items, and anything considered too detailed for the average undergraduate course has been ruthlessly cut from the text.

This book has not been written in isolation. Its style and content developed slowly as it was moulded into shape with the help and advice from colleagues and friends, all of whom gave their time freely. It gives me great pleasure finally to be in a position to thank them. I start with Ian Sherman, Blackwell Science Ltd (now at Oxford University Press), without whose enthusiasm and advice this book would never have happened; Nancy Dise, David Myrold, Peter Harris and Chris van Kessel for reviewing an early draft of the book. I also thank my colleagues at IACR-Rothamsted and Silsoe Research Institute, including Prof. David Powlson, Prof. David Jenkinson, Dr Keith Goulding, Paul Hargreaves, Dr Toby Willison, Dr Laurence Blake, Maureen Birdsey, Dr Phil Brookes, Dr Penny Hirsh, Dr Richard Webster, Jamie Allen, Dr Saran Sohi, Pete Falloon, Sharon Fortune, Pete Redfern, Tim Mauchline, Chris Watts, Dr Paul Hallett, Prof. Steve McGrath. A particular source of advice has been Prof. John Catt, who made numerous helpful suggestions. Other help, both practical and intellectual, was provided by Jason Ashman, Jo and Chris Rathbone, May and Bob Ashman, and Sarah Shannon at Blackwell Science Ltd.

A very special thank you must go to Dominique Niesten, who made a major contribution to this book. Her strong sense of chapter structure and clear writing style proved invaluable when revising the text. Without her advice and encouragement the manuscript would have never left my desk. Of course, all responsibility for errors of fact and interpretation remains with the authors.

M. R. Ashman, Rothamsted

List of Abbreviations

ANOVA   analysis of variance
CEC   cation exchange capacity
EC   electrical conductivity
FAO   (UN) Food and Agriculture Organization
FYM   farmyard manure
GIS   geographical information system
ICRCL   Interdepartmental Committee on the Redevelopment of
            Contaminated Land
ISSS   International Society of Soil Science
LR   leaching requirement
MAFF   Ministry of Agriculture, Fisheries and Food
MIT   mineralization and immobilization turnover
NV   neutralizing value
OC   organochlorine
PAH   polyaromatic hydrocarbon
PCB   polychlorinated biphenyl
rRNA   ribosomal ribonucleic acid
SMD   soil moisture deficit
SNS   soil nitrogen supply
SOM   soil organic matter
SOTER   Soils and Terrain Digital Data Base Project
Unesco   United Nations Educational, Scientific and Cultural
            Organization
USDA   United States Department of Agriculture
USEPA   United States Environmental Protection Agency
WRB   World Reference Base

1 Rocks to Soil

Introduction

Let's start with a question – what is soil and how does it form? Immediately you will know that soils are brownish, turn muddy when wet, and have some importance to plant growth. However, this level of understanding is insufficient when we really want to understand and appreciate the formation and utilization of soils for agricultural purposes. We can start by saying that rocks are transformed into soil by physical and chemical changes that occur at the Earth's surface. Gradually, these mineral inputs are combined with organic matter. Over time both the mineral material and organic matter are transformed into new materials; these are then moved through the soil by percolating water, so that the more soluble compounds are finally lost completely. It is the nature of these inputs, transformations, movements and losses that determines what type of soil will form. This chapter will look at four main lines of enquiry:

1 **Soil formation – what are the initial inputs?**
    What are soils made from?
    What are the main mineral inputs?
    Why do some rocks break up?
    How do rocks break up?
    What are the main organic inputs?
    What are plants made from?
    How does organic matter begin to accumulate in soil?
    How does soil organic matter differ depending on environmental
        conditions?

2 **How are these inputs transformed into new compounds?**
    What kinds of mineral material do soils contain?
    What are clay minerals and how do they form?
    How is organic matter transformed into new material?
    What is humus and how does it form?

3 **How is material moved and then finally lost from the soil?**
    How is soil material moved to new geographical locations?
    How is material moved within the soil?

4   How can we explain soil formation?
      Example one: podzol (spodosol)
      Example two: gleysol (aquepts)
      Example three: ferralsol (oxisol)
      Example four: histosol (histosol)

1   Soil formation – what are the initial inputs?

What are soils made from?

The first step in soil formation happens when mineral material from rocks and organic matter from plants and animals are combined together. Rock fragments without organic matter are unable to support plant growth. For example, think about sowing wheat seeds in gravel – what do you think your chances of a successful crop would be? Although organic matter without mineral inputs can support plant growth, composted organic matter lacks many of the physical characteristics that are commonly associated with soil. It is the combination of mineral and organic matter that gives soil its unique properties. Together they make up approximately 50% of the soil volume: the remaining 50% is pore space, filled with either air or water depending on how wet the soil is. We will start by looking at the main mineral inputs.

What are the main mineral inputs?

Rocks are composed of one or a number of different minerals. Geologists have traditionally divided rocks into three broad classes: igneous (formed from molten magma); metamorphic (rocks that have been altered by heat and pressure); and sedimentary rocks (formed from sediment deposits). Rocks differ from each other because they contain either different types of minerals or the same minerals but in varying quantities. This concept is not difficult to understand if you compare, let's say, natural rock and artificial building materials. We can produce a wide range of bricks, blocks and slabs, all with differing properties, by simply varying the proportions of sand, gravel and cement from which they are made: rocks are no different. Rocks with differing mineral compositions vary in their ability to withstand the natural forces that slowly disintegrate them. However, before looking at *how* rocks disintegrate, we should ask *why* they break up in the first place.

Why do some rocks break up?

The Earth's surface is made up of many different types of rock. Some, such as carbonate rocks like limestone and chalk, are composed of prehistoric marine creatures, whereas others like coal are derived from prehistoric plants. However, in most cases rocks are mainly composed of the element silicon. In rocks, silicon is usually combined with oxygen to form silica and silicates. One important difference between rocks is the amount of silica they contain. The proportion of silica to other minerals affects rock suscept-ibility to disintegration when exposed at the Earth's surface. Generally, for a given set of climatic conditions, the rocks with the lowest concentration of silica break up more quickly than those with higher concentrations. These differences in silica content occur as the molten magma solidifies to form rock. As the magma starts to cool, minerals with the lowest concen-trations of silica form first. They are followed by minerals with increasing silica contents. We can use the amount of silica igneous rock contains as a useful method of classification:

♦ ultrabasic: rocks with less than 45% silica, such as serpentinite and peridotite;
♦ basic: rocks with 45–55% silica, such as basalt, gabbro and dolerite;
♦ intermediate: rocks with 55–65% silica, such as amphibolite and andesite;
♦ acidic: rocks with over 65–85% silica, such as granite and pegmatite.

How do rocks break up?

A more scientific term for the break-up of rocks is 'weathering'. We use this term because climate, and the prevailing weather, is the main factor that eventually transforms rock to soil. Weathering can be either physical or chemical. Figure 1.1 shows how we can divide physical weathering into a number of separate processes.

Physical weathering

Thermal weathering   Different minerals have different rates of expansion when heated. When rocks composed of several different minerals are exposed to heat they experience different rates of expansion. This causes stress within the rock, which can result in fracturing in areas of weakness. Thermal stress can also be caused by temperature differences between the outer and inner parts of the rock. This form of thermal stress is called exfoliation or 'onion skin' weathering, because as outer layers of the rock are fractured, they are gradually peeled away.

Mechanical weathering   Water can penetrate rocks along small cracks. When water freezes, its volume expands by 10% (think about how uninsulated water pipes can burst in the winter). The force exerted by expanding ice is enough to break open cracks. This form of weathering is called 'frost shattering'. In addition, when exposed to water, different minerals often have different rates of swelling and shrinkage; this can initiate stresses within the rock that eventually cause it to fracture.

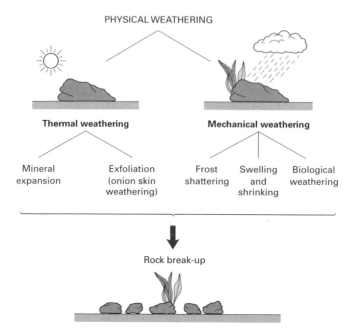

Fig. 1.1 The physical break-up of rocks by thermal and mechanical means.

Fig. 1.2 Plants can speed up physical and chemical weathering by expanding into rock fissures and excreting substances called exudates. These are then metabolized by micro-organisms, thereby increasing the rate of chemical weathering.

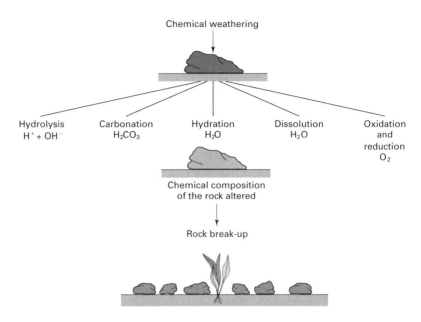

Fig. 1.3 The chemical break-up of rocks by hydrolysis, carbonation, hydration, dissolution, oxidation and reduction. Unlike physical weathering, which simply breaks the rock into smaller and smaller fragments, chemical weathering can also change the physical and chemical properties of the rock.

Another form of mechanical weathering comes from the pressure plant roots exert as they expand into crevices and small cracks in the rock. By widening existing cracks, more of the rock surface is exposed to the elements. This accelerates the weathering process and hastens the disintegration of the rock. Figure 1.2 is a good example of this process (it shows a small tree growing out of a crack in a rock).

Chemical weathering

In addition to being exposed to mechanical disruption rocks are also attacked chemically. Chemical weathering can be divided into the following set of processes (see Fig. 1.3).

Hydrolysis   This is the most common form of chemical weathering. It occurs when water molecules (H_2O) separate (dissociate) into two charged particles, H^+ (a hydrogen ion) and OH^- (a hydroxyl ion). The term 'ion' refers to the fact that the particle carries a charge. Hydrogen and hydroxyl ions attack the bonds that hold minerals together. Hydrolysis not only causes rock disintegration but it also changes the chemical nature of the minerals. Hydrolysis is a very important process in soils and it is essential you understand the mechanism: it is shown in Fig. 1.4.

For **cations** (positively charged ions)

$$M^+ + H_2O \rightarrow MOH + H^+$$

For **anions** (negatively charged ions)

$$X^- + H_2O \rightarrow HX + OH^-$$

Fig. 1.4 Hydrolysis is the chemical reaction of a compound with water. The chemical reactions for cations and anions are shown.

$$CO_2 + H_2O \rightarrow H_2CO_3 \rightarrow H^+ + HCO_3^- \rightarrow 2H^+ + CO_3^{2-}$$

| Soil respiration | Soil water | Formation of acid | Dissociation | Further dissociation |

Fig. 1.5 Biological activity in the soil leads to carbon dioxide being respired. This dissolves in soil water to produce a weak acid that can then attack minerals.

Carbonation   This is an accelerated form of hydrolysis, which is caused by biological activity within the soil. The majority of soil organisms respire carbon dioxide (CO_2). When CO_2 comes into contact with water, a proportion of it dissolves to form carbonic acid (rain is naturally acidic for this reason). Plant roots are particularly destructive in this respect because, in addition to carbonic acid produced by respiration, they also excrete sugars that are then used and converted to acids by micro-organisms in a process not unlike tooth decay following a prolonged sugary diet. All acids are rich sources of hydrogen ions; carbonation therefore enhances hydrolysis. The process is shown in Fig. 1.5.

Hydration   During hydration minerals absorb water, but unlike hydrolysis there is no ion formation: during hydration the water molecule remains intact. When a mineral undergoes hydration its physical and chemical properties can be altered. A similar process occurs when pasta is immersed in water: think about how physical characteristics are altered as it absorbs water. When some minerals become hydrated they can also become weakened physically.

Dissolution   In this process minerals simply dissolve in water. A few minerals such as sodium chloride (table salt) and potassium chloride are completely soluble in water. Minerals such as these dissolve, and are then washed away in solution.

Oxidation and reduction   When exposed to the atmosphere some minerals undergo chemical changes; some are 'oxidized' and others are 'reduced'. In its simplest form oxidation can be regarded as a mineral's tendency to take up oxygen, while reduction is its ability to lose oxygen. However,

Table 1.1 Broad soil carbon and nitrogen ratings.

Rating	Carbon (%)	Nitrogen (%)
Very high	>20	>1.0
High	10–20	0.5–1.0
Medium	4–10	0.2–0.5
Low	2–4	0.1–0.2
Very low	<2	<0.1

Source: J. R. Landon (1984) *Booker Tropical Soil Manual*, Longman Scientific and Technical, Harlow.

this narrow definition has been expanded so that it also refers to the loss (oxidation) or gain (reduction) of electrons. Although chemically the process can become quite complicated, the main point is that changes in a mineral's oxidation state can weaken it.

Although it has been convenient to subdivide weathering into chemical and physical processes, you must appreciate that they are not mutually exclusive. For example, the break-up of a large rock into smaller fragments by physical weathering also increases its rate of chemical weathering by exposing a greater proportion of its surface to the prevailing weather conditions. Put simply, in most situations smaller fragments of rock have higher rates of chemical weathering.

What are the main organic inputs?

Weathered rock finally becomes soil when organic matter is incorporated with mineral fragments. The term 'soil organic matter', in its widest sense, covers all the living and dead organisms contained within the soil. However, when soil scientists use the term they are usually referring to the remains of plants and animals. In some cases these residues are recent additions to the soil, while others may be many years old. When measuring soil organic matter soil scientists often refer to the concentration of soil carbon, because carbon is the main constituent of organic matter, typically accounting for around 58% of the total weight. Table 1.1 shows some typical soil carbon and nitrogen concentrations.

Soil organic matter not only supplies nutrients to plants but it also alters the physical nature of the soil by binding soil particles together into discrete units called aggregates (see Chapter 2). Before looking at soil organic matter in more detail, we will consider briefly what plant material is made from, because in most soils plant material is the commonest type of organic input.

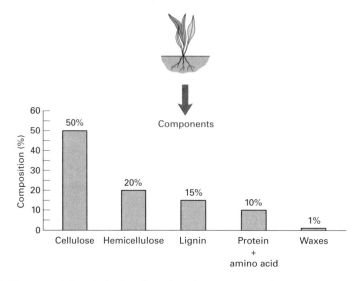

Fig. 1.6 Plant material can be broken down into the components shown.

What are plants made from?

In many respects the breakdown of soil organic matter is similar to rock weathering. As with rocks, the relative proportion of resistant to degradable compounds will largely determine how quickly plant residues are broken down by soil micro-organisms. When we considered rock break-up we used the term 'weathering'; when plant and animal residues are reduced to simple chemicals we use the term 'mineralization'. The main compounds present in plant material are shown in Fig. 1.6.

The main compounds

Cellulose and hemicellulose   Plant tissue consists mostly of cellulose (15–50%) and hemicellulose (10–30%). Plant walls are made from a combination of cellulose fibres that are encrusted with hemicelluloses. Both compounds are made up of sugar molecules joined together to form long chains.

Proteins and amino acids   Proteins and amino acids also consist of carbon compounds, but unlike cellulose they also contain considerable quantities of nitrogen. Typically, plant material consists of approximately 5% protein and 5% amino acids. Nitrogen is a very important nutrient and is essential to all living organisms. Plant material that is rich in protein therefore offers a valuable food source for soil micro-organisms.

Lignin   Whereas proteins are attacked quickly, lignin is only mineralized slowly. It is the most resistant compound found in plant material. Plant material consists of approximately 15% lignin, whereas woody plant tissues contain higher concentrations, approximately 25–30%. The mineralization of woody tissue is undertaken by specialist organisms, mainly fungi, which slowly mineralize lignified material. Just as resistant mineral material can accumulate in the soil, so can resistant organic compounds such as lignin.

Fats and waxes   Although these compounds make up only a small fraction of the dry-matter weight of deciduous plant material (1–5%), their concentration in coniferous plant litter may be as high as 20–25%.

How does organic matter begin to accumulate in soil?

Before soil organic matter can begin to accumulate in the soil, mineral particles need to be colonized by plants: this presents a problem. Rock fragments alone do not offer the most attractive conditions for a colonizing plant because they lack the ability to supply adequate quantities of water and nutrients. However, certain bacteria, fungi and plant species have evolved to live in water- and nutrient-limited environments. These organisms are called 'primary colonizers'.

In many cases primary colonizers are able to obtain nutrition from sources other than the soil. This allows them to live in areas that other organisms would find too hostile. One important adaptation is the ability to obtain nitrogen (one of the most important plant nutrients) from atmospheric rather than soil sources. Organisms that have the ability to do this are called 'nitrogen fixers'. One example of a primary colonizer that can fix atmospheric N is lichen.

Lichens are a mutually beneficial association between algae and fungi. The algae obtain carbon and in some cases nitrogen from the atmosphere using a combination of photosynthesis and N-fixation. Once their own carbon and nitrogen requirements have been met, surplus nutrients are then passed to the fungi. For their part, the fungi attack the rock with organic acids. This releases minerals for the algae. This type of biological process, whereby both organisms benefit by forming a close association with each other, is referred to as 'symbiosis'. When these specialized organisms die, their tissues become combined with the mineral material, so forming the first organic-matter additions to the soil. However, it is important to note that not all N-fixers are primary colonists and not every primary colonizer can fix nitrogen. Other adaptations, such as the ability to obtain mineral nutrition from rock surfaces, also play an important part in the success of many primary colonizers.

As the amount of soil organic matter increases it becomes possible for other plants (which extract their N from soil, rather than atmospheric sources) to colonize the site. When these plants die, their residues are then

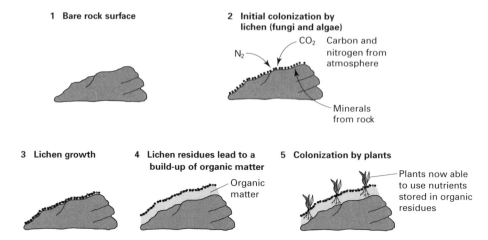

Fig. 1.7 One example of primary colonization is the role lichen plays in the initial stages of soil formation.

incorporated into the store of soil organic matter. The processes involved in primary colonization and soil formation are shown in Fig. 1.7.

How does soil organic matter differ depending on environmental conditions?

As ecosystems mature they develop their own characteristic vegetation in response to local environmental conditions. This means that soils supporting different types of vegetation will also contain different types of organic matter. For example, in some upland areas of the UK the soils are acidic and so support vegetation characteristic of acidic environments. Inputs of organic residues from acid-tolerant vegetation such as Ericaceae are relatively resistant to microbial attack because of their high concentrations of lignin. Lower levels of microbial activity, coupled with organic matter that is difficult to decompose, can lead to the accumulation of plant residues in a black-coloured layer in the top few centimetres of the soil. In neutral to alkaline soils, organic matter tends not to accumulate because high levels of biological activity and mixing by soil animals, such as earthworms, lead to a more uniform distribution of the organic matter throughout the whole soil. These two contrasting soil processes are normally visible in the field and are often associated with certain soil types. We will look at these processes in more detail later in this chapter.

So far we have looked at the main mineral and organic matter inputs, but before going further let's imagine mixing rock fragments with chopped-up plant residues – would the mixture resemble a soil? Probably not, and the reason why it doesn't is that the mixture still lacks some very important material. Although we have described the initial mineral and organic inputs, we now need to consider how they are transformed into new compounds. This is because weathering and mineralization do not simply break down mineral and organic material into smaller units, but transforms them, both chemically and physically. It is these transformed compounds that give soil many of its unique properties.

Essential points

♦ Rocks break up as they adjust to new conditions at the Earth's surface.
♦ Weathering is an umbrella term that covers a whole range of processes, all of which result in rock disintegration. It can broadly be divided into physical and chemical processes.
♦ Soils differ from weathered rock fragments because they contain an organic component, made up of living and dead organisms.
♦ Plant material is made up of five fractions: cellulose and hemicellulose, proteins, lignin, fats and waxes. These fractions differ in the ease with which they are mineralized by soil micro-organisms.

2   How are these inputs transformed into new compounds?

What kinds of mineral material do soils contain?

We can fractionate the mineral component of a soil into three particle-size classes: these are sand, silt and clay and their sizes are shown in Fig. 1.8.

The sand fraction (largely composed of resistant quartz grains) has the largest particles, 0.06–2 mm (or 0.05–2 mm in the US system and 0.02–2 mm in the International System) and the clay fraction the smallest. In fact, clay particles are so small (2 μm) that we use a special symbol 'μ' to describe their size. One μ is simply one millionth (1/1 000 000) of a metre. Apart from their size, sand and silt particles are largely unaltered by chemical weathering. Chemically they are the same as the mineral material in the parent rock, mainly composed of resistant minerals such as quartz. Clay-sized particles, on the other hand, are referred to as 'secondary minerals' because, unlike primary minerals, they have already undergone one phase of chemical weathering, which has altered their physical and chemical composition. Within the clay size range there are several different minerals; we will concentrate on the aluminosilicate clay minerals.

What are clay minerals and how do they form?

The combined effect of hydrolysis, hydration and dissolution causes rocks to disintegrate into their chemical constituents, mainly silicon, iron, aluminium, magnesium, potassium and calcium. Once these elements are released from the old skeletal framework of the primary mineral they are able to

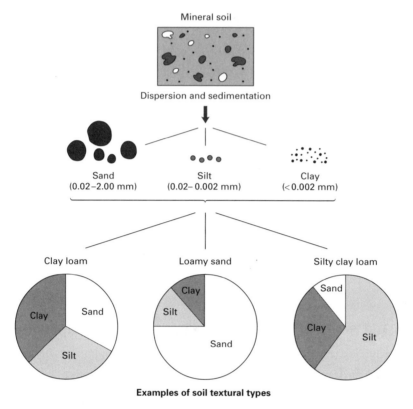

Fig. 1.8 Soil can be fractionated into mineral fractions. The amount of sand, silt and clay in a soil will determine its texture. There are many different textural classifications. Here we have shown three: clay loam, loamy sand and a silty clay loam.

recrystallize to form secondary clay minerals (Fig. 1.9). The ratio of silica to the other elements largely determines which type of clay mineral will form.

The structure of clay minerals can be complex. However, we can simplify things by saying that all clay minerals are composed of sheets of interlocking silica that alternate with sheets of aluminium oxide. We can use differences in the number of silica to aluminium sheets as a way of dividing them into classes. This makes good sense because clay minerals with different structures (as determined by the silica to aluminium ratio) have slightly different physical and chemical characteristics (this topic is dealt with in greater detail in Chapter 3).

How is organic matter transformed into new material?

It is not only minerals that undergo chemical changes within the soil. Organic residues are altered to new material called 'humus'. Humus is both chemically and physically different from the chemicals that make up plant and animal material.

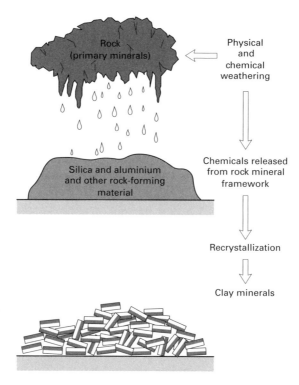

Fig. 1.9 A simplified version of clay mineral formation.

What is humus and how does it form?

Unlike clay minerals, humus is very difficult to classify because its composition is so variable. It can be regarded as something like a chemical 'junk yard' because it is a repository for resistant plant material and microbial waste products. These combine to form very complicated molecules in a process called 'humification'. Figure 1.10 shows a simplified model of humus formation.

Compared with plant residues, humus is very stable. It is only slowly broken down and mineralized to its constituent chemicals, many of which are valuable plant nutrients. Unlike plant residues that only persist in the soil for a few years, some of the organic carbon in humus can be over 1000 years old. As humus becomes intimately mixed with the mineral fraction it imparts a dark brown colour to the soil.

Now, let us return to the hypothetical experiment that we described earlier in this section, when we unsuccessfully attempted to create soil by mixing chopped-up plant material and finely crushed rock fragments. If, in addition to these, we take clay minerals and humus and mix these together, we finally start to create a mixture that resembles soil. However, soils are not static; they change over time, so in order to complete this

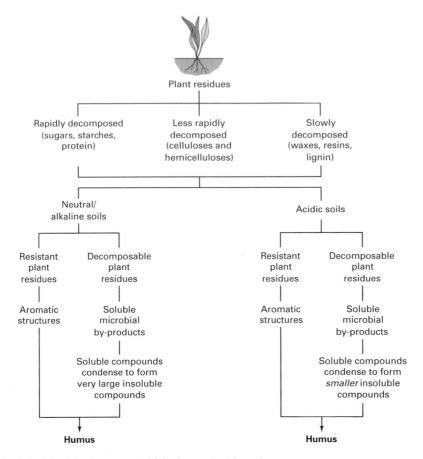

Fig. 1.10 Humification in neutral/alkaline and acidic soils.

picture of soil formation we need to look at one last set of processes: how soil components are moved and lost from the soil.

Essential points

♦ Initial inputs of mineral and organic matter are transformed into new compounds, which have different physical and chemical properties.
♦ Primary minerals are converted to secondary minerals, the most important of which are the clay minerals.
♦ Plant residues are initially mineralized and recycled by soil microorganisms, ultimately producing humus.

3   How is material moved and then finally lost from the soil?

There are two important types of movement: the movement of fine mineral particles to new geographical locations and the movement of material within the soil.

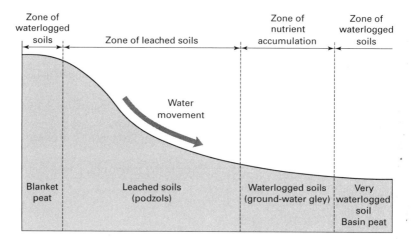

Fig. 1.11 A soil catena showing a typical sequence of soil types found in a temperate high-rainfall area.

How is soil material moved to new geographical locations?

Wind and water can sometimes move unconsolidated mineral particles to new geographical locations. The movement of material in this way explains why in some cases a soil's mineral characteristics bear little resemblance to those of the underlying rocks of the region. Soils can develop on material that has slipped down hill slopes (solifluction terraces), sediment material deposited by rivers (river terraces and flood plains) and fine silty wind-blown material (loess deposits). We will not consider these further as they can be regarded simply as a type of soil input which is subjected to all the chemical and biological transformations we have discussed already.

In some cases soil chemicals can move down-slope, producing different soil types depending on their location. This is despite the fact that all the soils of the area have similar inputs. The reason for this is that water moving from upper to lower slopes transports material down-slope, so that soils on the upper slopes can become thin and depleted of material while those down-slope become enriched with material. This process can lead to a well-defined sequence of soil types relative to their location on a slope, sometimes referred to as a 'soil catena' (Fig. 1.11).

How is material moved within the soil?

The other type of movement occurs vertically and horizontally within the soil. When soil scientists talk about soils they sometimes use the term 'profile'. The soil profile not only refers to the top layers of soil, but also includes all the underlying layers down to the unaltered parent material on which the soil has formed. The movement of mineral and organic material

Table 1.2 Master horizons and suffixes.

Master horizons

O	Upper organic horizon consisting of peaty material. Often found under wet conditions
A	Surface horizon in which organic matter and mineral material are closely associated
E	Horizon below the A and O, which is characterized by being depleted of material
B	Mineral horizon that is characterized by being a zone of accumulation of organic matter, minerals and inorganic chemicals
C	Horizon of unconsolidated parent material
CR	An intensely gleyed horizon
R	Parent bedrock

Suffixes (partial list)

a	ashen-coloured horizon
ca	accumulation of calcium carbonate
fe	illuvial concentration of iron
g	strong mottling due to periodic wetness
h	humified well-decomposed organic matter
s	a B horizon enriched with aluminium and iron oxides
t	accumulation of clay
w	a horizon showing alteration due to weathering

---

down the profile is determined by how easily the material moves in water and the rate of water movement. Mobile compounds move downwards, from the upper to lower layers of the soil, so that some parts of the soil profile become enriched while others are depleted of material. The movement of chemicals within the profile, at varying speeds, produces a number of distinct layers. These are referred to as 'soil horizons'.

Soil scientists who specialize in studying the formation and classification of different soil types are called 'pedologists'. In order to describe soils in a systematic way, pedologists have labelled the horizons with increasing depth from 'O' through to 'R' (see Table 1.2). Figure 1.12 shows how these horizons are arranged within a hypothetical soil profile.

Under acid conditions the upper O horizon is sometimes visible as a dark-coloured band which can be further divided into three separate layers referred to as L (litter), F (fermentation) and H (humic). Organic matter in this form is described as 'mor'. The other extreme is under neutral-to-alkaline soil conditions, where a combination of readily mineralized residues and high numbers of soil organisms rapidly break down the organic matter and mix it with the underlying mineral layers. Under these conditions there is often no distinctive black-coloured O horizon and no L,

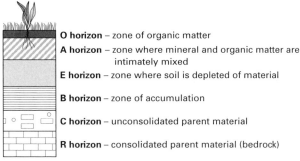

O horizon – zone of organic matter
A horizon – zone where mineral and organic matter are
                    intimately mixed
E horizon – zone where soil is depleted of material

B horizon – zone of accumulation

C horizon – unconsolidated parent material

R horizon – consolidated parent material (bedrock)

(NB: not all soils have all these horizons)

Fig. 1.12 The soil profile and its subdivision into soil horizons. The presence or absence of certain horizons is used to group soils into classes.

F and H subhorizons. When organic matter is in this form it is described as 'mull'. Soil profiles with both mull and mor characteristics are referred to as 'moder' (see Fig. 1.13).

Referring back to Fig. 1.12, the A horizon is characterized by having high concentrations of organic matter. It is sometimes underlain by an E or an 'eluvial' horizon, which means that this part of the soil profile has a tendency to lose material. This transported material tends to be deposited lower down the soil profile in a zone of deposition, which is termed the B horizon or 'illuvial' horizon. If you find it hard to remember this terminology, just remember eluvial has 'e' for 'exit' and illuvial has 'i' for 'in'. The C horizon is a zone of unconsolidated mineral material that lies directly above the consolidated rock (R), often referred to as the 'parent material'.

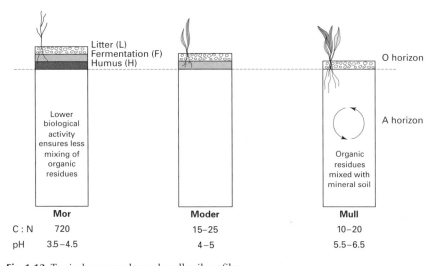

	Mor	Moder	Mull
C : N	720	15–25	10–20
pH	3.5–4.5	4–5	5.5–6.5

Fig. 1.13 Typical mor, moder and mull soil profiles.

Although we said mobile substances move down soil profiles, creating zones of depletion and accumulation, there are subtle differences in the way substances move. Movement can be in either solution or suspension. If you are unclear about the difference, think about what happens when you add a tablespoon of sugar to a glass of water: the sugar completely dissolves, i.e. it goes into solution. In other words, we could not filter the sugar out of the water. Now tip in a tablespoon of flour and stir vigorously: the flour is held briefly in suspension, but does not dissolve. If we needed to we could extract it from the water quite easily by filtering or leaving it to settle. In a soil, material can move down the profile in solution (like the sugar) or suspension (like the flour). We refer to these processes as 'leaching' (solution) and 'eluviation' (suspension). Both can occur simultaneously as water moves through the soil profile.

Leaching involves the movement of soluble ions such as Ca^{2+}, Mg^{2+}, Na^+, K^+, NO_3^-, NO_2^- and complex anions based on SiO_4^{4+}. In the case of iron (Fe), its solubility depends on whether it is in an oxidized (Fe^{3+}) or reduced (Fe^{2+}) form. Distinctive colour changes from red to grey enable soil surveyors to identify zones where the iron is in an oxidized or reduced state (more on this later). As leaching continues, the more soluble compounds are completely washed out of the soil profile. Under conditions of extreme and prolonged weathering such as those found in the humid tropics, clay minerals are completely weathered away, leaving oxides of aluminium (Al) and iron (Fe).

Eluviation is the movement of insoluble particles such as clay minerals down the profile. This process is sometimes called 'lessivage'. In the field, lessivage is detectable by a relative increase in clay content of the B horizon when compared with either the A or E horizon. Lessivage is sometimes detectable in the field by the presence of thin coatings of clay on the surfaces of soil aggregates; these are referred to as 'clay skins'.

One of the main tasks of the pedologist is to piece together the sequence of horizons in the soil profile in order to explain how a particular soil has formed. The type and location of horizons can be used as a way of grouping similar soils together into classes. This topic will be covered in more detail in Chapter 5.

Essential points

♦ Soils are not inert but dynamic: as new mineral and organic material is incorporated, old material is broken down, moved and lost from the profile.
♦ The downward movement of material by leaching and eluviation creates a number of distinct layers within the profile called horizons.
♦ Horizons can provide important clues about the major inputs, transformations, movements and losses from the soil.

4   How can we explain soil formation?

The type of mineral and organic inputs, and the speed at which they are transformed, moved and then finally lost from the soil profile, will depend on local environmental conditions. As the soil develops, its profile will slowly change. The famous Russian soil scientist Vasily Dokuchaiev (1846–1903) made a number of major contributions to our understanding of soil development by showing how climate, bedrock and organisms interact to produce different kinds of soil. In 1941 Hans Jenny extended Dokuchaiev's original ideas to give five soil-forming factors, which are: climate, parent material, topography, organisms and time. Sometimes these factors are expressed in a similar way to a mathematical equation.

Soil formation = climate, parent material, topography, organisms (plants and animals), time.

These five factors determine the nature of the initial inputs, how they are transformed and how quickly they are moved and lost from the soil. Because soils form under a range of conditions, it is not surprising that so many different soil types are found throughout the world. In the final section of this chapter we will describe four soil profiles. Although they represent only a small fraction of the many soil profiles that have been recorded by pedologists, they do provide a good illustration of some of the processes that we have already discussed. Here we have used their popular names and give their US name in parenthesis.

Example one: podzol (spodosol)

Profile description: These soils are widely distributed across the cool, humid and temperate regions of the world. They have very distinctive profiles with striking differences between the horizons. The profile consists of a relatively deep O horizon, consisting of plant material, which can sometimes be subdivided into L, F and H subhorizons. These horizons contain plant material in varying stages of decay. Fungal mycelium and small arthropods are normally the dominant soil organisms. The A horizon has a dark grey/black colour, commonly containing 10% organic matter, beneath which lies a bleached, light-coloured, E horizon (which is labelled 'Ea' in Fig. 1.14). The B horizon is a zone of accumulation. The movement of iron is often revealed by a colour change from light brown/grey to red. When the environment is very acidic and the rainfall high, the iron may form a thin red horizon (Bfe) known as an 'iron pan'. Sometimes this is overlain by a thin dark organic layer (Bh).

Inputs: Rainfall is critical in the formation of podzols. In lowland areas of the UK where rainfall is generally 600–700 mm per year, podzols are

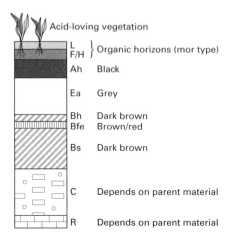

Acid-loving vegetation

L
F/H } Organic horizons (mor type)

Ah   Black

Ea   Grey

Bh   Dark brown
Bfe   Brown/red

Bs   Dark brown

C   Depends on parent material

R   Depends on parent material

Fig. 1.14 Podzol (spodosol).

usually confined to coarse-grained sands. However, in areas that experience heavy rainfall, such as Western Scotland and Norway, podzols can be found in areas containing basic mineral material. Podzol formation is also linked with acid-tolerant vegetation. These plants produce resistant litter that is only mineralized slowly.

Transformations: Organic acids are produced as a by-product from the mineralization of plant litter. These lead to high rates of chemical weathering within the profile. As clay minerals break down they release silica, aluminium and iron. Resistant plant residues and a soil microbial population dominated by fungi lead to the formation of mor humus.

Movements and losses: Podzol development is highly dependent on high rates of rainfall and rapid free drainage. Iron and aluminium, released during the break-up of clay minerals, move down the soil profile before collecting in the B horizon. This creates a bleached E horizon and a red-coloured B horizon. Under very wet, acidic conditions organic matter in the L, F and H horizons may also move down the soil profile before collecting as a dark-coloured band just above the iron pan.

Example two: gleysol (aquepts)

Profile description: Many soils in the northern latitudes suffer from periodic waterlogging. Soils that have horizons that show signs of periodic waterlogging are called gleysols (aquepts) or simply gley soils. The water can be from either above ground (surface water) or below ground (ground water sources). Waterlogging from surface-water sources occurs when drainage is impaired by compacted lower horizons. Inputs of water from either rainfall or river flooding only manage to percolate through the soil slowly, so that the surface horizons remain waterlogged. Soils showing these characteristics

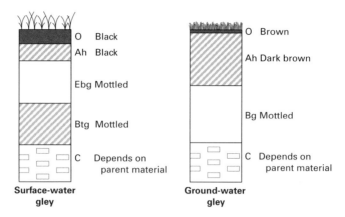

Fig. 1.15 Gleysols (aquepts).

are referred to as 'surface-water gleys'. In the case of 'ground-water gleys', it is the lower horizons that are periodically waterlogged, usually as a result of a high water table. Waterlogged horizons are visible by colour changes, as brown shades become progressively grey-coloured. These grey-coloured horizons are often punctuated by red/brown mottles, especially along root channels and large cracks (see Fig. 1.15).

Inputs: The most important input that leads to the development of gleysols is water. Mineral and organic matter can be from a variety of sources because gley features can be detected in many different soil types. The United States Department of Agriculture (USDA) system of soil classification, 'Soil Taxonomy', does not recognize gleysols as a separate classification order. Gleying is seen more as a process, rather than as a distinct soil type, because it can be detected in many soils that experience periodic waterlogging.

Transformations: The movement of oxygen through water is much slower than through air. When a soil becomes waterlogged its microbial community (most of which needs oxygen) experiences oxygen starvation (anaerobic conditions). However, unlike higher organisms, some groups of soil microbes can use alternative forms of respiration whereby other compounds, such as iron (Fe^{3+}), are used instead of oxygen. During this process the chemical nature of the iron is transformed from Fe^{3+} to Fe^{2+}. In other words, it is reduced. Sometimes the two different forms of iron are referred to as ferric (Fe^{3+}) and ferrous (Fe^{2+}). This process is visible by a colour change within the soil profile from red to grey, because aerobic iron (Fe^{3+}) is characterized by a reddish colour while anaerobic iron produces grey-coloured horizons. When oxygen penetrates the waterlogged soil along root channels and large cracks, iron in the reduced Fe^{2+} form is converted back to the oxidized Fe^{3+} form, causing red-coloured mottling.

Movements and losses: The water solubility of the two forms of iron varies. When iron is in its anaerobic (Fe^{2+}) form it is more water-soluble. This allows it to move down the soil profile so that it accumulates in the lower horizons.

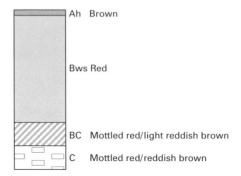

Ah   Brown

Bws Red

BC   Mottled red/light reddish brown

C   Mottled red/reddish brown

Fig. 1.16 Ferralsol (oxisol).

Example three: ferralsol (oxisol)

Profile description: The wet humid tropics form a band that lies either side of the equator and are characterized by high temperatures and rainfall. Ferralsols are often found in this region and occupy large areas of central Africa and South America. They are characterized by often having a loose, thin covering of plant litter, underlain by a greyish-brown A horizon that changes abruptly to a deep, bright red-coloured B horizon (Fig. 1.16). This horizon normally contains very low concentrations of organic matter (1%) and primary minerals that are susceptible to weathering.

Inputs: The combination of high rainfall rates and high temperatures over several million years has created an intensely weathered soil profile. Although these soils may support lush surface vegetation, such as rain forests, the concentration of soil organic matter is generally quite low, ranging from 1% to 3% in the A horizon. This is because rapid rates of mineralization lead to the loss of even the most resistant components of the plant material and humic materials contained in the soil organic matter.

Transformations: The intense weathering conditions lead to the rapid break-up of minerals releasing soluble silica, iron and aluminium. More insoluble compounds such as the oxides and hydroxides of iron and aluminium (sometimes referred to as goethite and gibbsite) persist within the profile. Primary minerals that are susceptible to weathering are sometimes completely absent from the profile. If the silica content of the parent material is high, clay minerals with simple structures can form, but when the mineral inputs are deficient in silica the profile can become completely dominated by the products of advanced weathering, mainly iron and aluminium oxides. Mineralization of organic matter is rapid, so that soil organic matter concentrations remain relatively low. A combination of iron oxides and low organic matter concentrations give these soils their distinctive red colour.

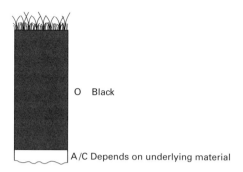

O   Black

A/C Depends on underlying material

Fig. 1.17 Histosol (histosol).

Movements and losses: One of the main mineral losses from ferralsols is the loss of silica, as mineral material is completely weathered in a process called 'desilication'. In areas where the topography is undulating, dissolved silica is leached down-slope. This leaves ferralsols on the upper slopes dominated by the oxides of iron and aluminium, and other soils on the valley floor enriched with silica. The influx of dissolved silica down-slope is often detected by higher concentrations of clay minerals in the down-slope soils.

Example four: histosol (histosol)

Profile description: Histosols are commonly known as peats. They cover large areas of Northern Europe, Canada and Russia, where there is a combination of high vegetative growth and persistent waterlogged conditions. Histosols can form in waterlogged depressions (basin histosols) or in areas where rainfall and humidity remain high throughout the year (blanket histosols). In acidic environments the profile is almost completely made up of organic matter, mainly from sphagnum mosses, moor grass and heather. Microbial activity is very low, which allows the rapid accumulation of organic-matter residues. The upper O horizon can be over 40 cm deep with only very subtle differences between horizons (Fig. 1.17).

Blanket and basin histosols vary in their chemical properties. The chemistry of basin histosols is heavily dependent on the chemical composition of the inflowing water. Generally basin histosols tend to be less acidic than blanket histosols.

Inputs: The most important inputs are water and a continuous supply of organic matter.

Transformations: All transformations are centred on the organic matter. Plant residues in the lower horizons tend to be more highly decomposed, so that there is a gradual transition from living tissue and fibrous plant residues in the upper horizons, to a highly humified, jelly-like substance in the lower horizons.

Movements and losses: As histosols are waterlogged for long periods, organic-matter decomposition is extremely slow. Soil organisms cannot completely decompose plant material and humified organic matter under anaerobic conditions. Movements and losses of material from the profile tend to be relatively low, although water draining from acidic blanket bogs tends to be stained brown with chemicals from the organic matter.

Essential points

♦ Varying inputs, transformations, movements and losses give rise to different soil types. Distinctive horizons record the slow evolution of a soil and allow us to piece together the main soil-forming factors.

♦ Podzols (spodosols): soils that form in areas with high rainfall and free-draining mineral material, usually with high concentrations of sand. Organic inputs are characteristically from acid-loving plants. As plant residues are transformed, mor-type organic matter is produced. Translocations lead to a bleached 'E' horizon and the formation of a rust-coloured 'B' horizon.

♦ Gleysols: the main defining feature in these soils is that they are subject to periodic waterlogging from ground water or surface water. This leads to the reduction of iron and its movement down the profile. This is visible when brown-coloured horizons give way to grey-coloured horizons that often contain red mottles along the main root channels.

♦ Ferralsols: the defining features of these soils are that they are old, with long histories of extreme weathering. Desilication and the gradual loss of primary minerals cause the soil to become progressively dominated by oxides. The main transformations are from primary minerals to residual oxides. Organic matter is mineralized quickly so that little accumulates. A combination of high rates of weathering and mineralization gives these soils their distinctive bright-red colour.

♦ Histosols: water and organic matter dominate inputs. Lack of oxygen slows mineralization and leads to the rapid accumulation of organic matter. Although the transformation of these residues is slow and incomplete, the deeper horizons tend to contain more highly humified material.

Chapter Summary

Soils are characterized by mineral and organic matter inputs. Some primary minerals are unstable at the Earth's surface: these minerals are susceptible to weathering and are transformed into secondary minerals. The most important of these secondary minerals are the silicate clays. Clay minerals have different physical and chemical properties to the parent material from which they originated. Plant and animal residues are subjected to microbial decomposition in a process called mineralization. During mineralization organic matter is reworked over and over again by micro-organisms, ultimately forming humus in a process called humification. Humus is only slowly mineralized. Soil material is moved within the profile by leaching and lessivage. These processes can lead to the formation of several distinctive layers in the soil profile; these are called horizons. The formation of well-defined soil types will depend on environmental variables such as climate, organisms, topography, parent material and time – these are known as the soil-forming factors. Soil profiles reveal the historical development of the soil by recording the main inputs, translocations and losses.

2 Particles, Structures and Water

Introduction

Imagine digging a hole – you lever out a wedge of soil, then turn it over, so that it falls in a heap. As you inspect the soil you might notice several important physical characteristics. For instance, after it is moistened does it feel pliable, so that it can be moulded into shapes, or does it feel gritty? Is it a collection of individual small structures bonded together or a structureless mass of individual particles? If the hole is dug 48 hours following heavy rainfall, does the soil remain wet and sticky or is it dry and brittle? This chapter will look at the importance of a soil's physical characteristics. It will look at four main lines of enquiry:

1 **What are the main soil particles?**
   What are soil particles?
   Why is soil texture such an important property?

2 **What is meant by the term 'soil structure'?**
   What is meant by 'soil structure'?
   How are soil aggregates formed?
   How are microaggregates formed and stabilized?
   How are macroaggregates formed and stabilized?
   How can we describe soil structure?

3 **How does water behave in the soil?**
   What is soil water?
   How does pore size affect water retention?
   What forces are acting upon the soil water in a saturated soil?
   What forces are acting upon the soil water in an unsaturated soil?

4 **Why are the physical characteristics of the soil so important?**
   Why is soil structure important?
   Why is soil water tension important?

1   What are the main soil particles?

What are soil particles?

Let us return to the hypothetical hole we dug at the beginning of this chapter and this time take one of the lumps of soil and slowly break it apart into several particle-size classes by sieving and sedimentation. If we were now to rank the mineral particles in order of decreasing size, at one end we might have large stones that fill the palm of your hand, and at the other individual particles so small they are invisible to the human eye. Given this vast size range, it is not surprising that soils with a high proportion of large particles will not have the same physical characteristics as those with greater amounts of small particles.

Soil mineral particles can be grouped together into five broad classes. These are shown in Table 2.1. First, particles such as stones, boulders and gravel larger than 2 mm are separated from soil material less than 2 mm using sieves. The soil that passes through the 2 mm sieve is called 'fine earth'. Fine earth is then divided into three particle-size fractions: sand, silt and clay. The proportion of sand, silt and clay in the fine earth is referred to as the soil's 'texture' and is used to classify soils into several textural groups.

Experienced soil scientists can identify a soil's textural class simply by rolling a lump of moistened soil between their fingers. A more accurate laboratory method can also be used, whereby a small sample of the soil is shaken with water to form a suspension (see Fig. 2.1). As the density of all the soil particles is similar (2.65 g/cm^3, which is the density of silica), the rate at which they settle is determined principally by their size. If samples of the suspension are taken at increasing time intervals, they will initially contain sand, silt and clay, then silt and clay, until finally, only the clay will be left in suspension. We can then place each sample in a watch glass and wait until the water evaporates. Once left with a dry residue we can weigh each sample. As we know that the first sample we

Table 2.1   The size range of particles in the UK, US and International System of classification.

Fraction	UK System	US System	International System
Stones/gravel	>2.0 mm	>2.0 mm	>2.0 mm
Coarse sand	2.0–0.2 mm	2.0–0.2 mm	2.0–0.2 mm
Fine sand	0.2–0.06 mm	0.2–0.05 mm	0.2–0.02 mm
Silt	60–2 µm	50–2 µm	20–2 µm
Clay	<2 µm	<2 µm	<2 µm

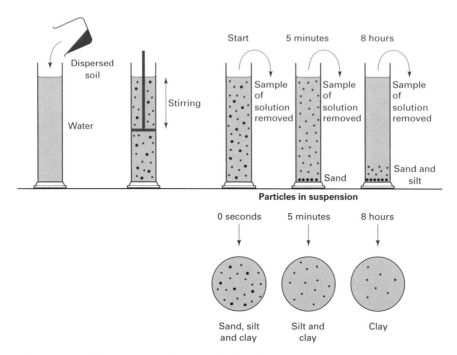

Fig. 2.1 Determining soil texture by mechanical analysis.

took contains sand, silt and clay and the second sample silt and clay, and the final sample just clay, we can calculate the percentage composition of each sample by difference. This method of textural analysis is called 'mechanical analysis'.

Once we have calculated the proportion of sand, silt and clay we can use the soil textural triangle shown in Fig. 2.2 to classify the soil. For example, if we found that a soil had 10% clay, 60% silt and 30% sand then it would be classified as a sandy loam in the UK system and a silt loam in the US system.

Why is soil texture such an important property?

There are two reasons why texture is a fundamental soil property. First, let us consider particle size because there is a 1000-fold difference in the size of the sand and clay particles. This means that the gaps between the sand particles tend to be large while those between individual clay particles are small. These gaps are called 'soil pores' and are important for many soil processes. We can illustrate the relationship between particle size and pore size using an everyday example, such as the gaps that would lie between a random collection of house bricks and a pile of dry builders' cement. Study Fig. 2.3; see how the gaps between the house bricks are large when compared with the gaps separating individual cement grains.

(a) **USDA** The soil textural triangle based on USDA size ranges, with broad groupings of textural classes

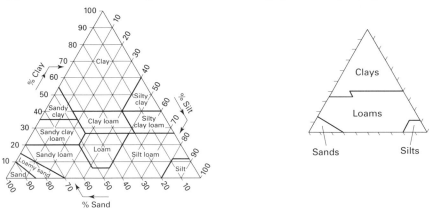

(b) **UK** The soil textural triangle based on Soil Survey of England and Wales size ranges, with broad groupings of textural classes

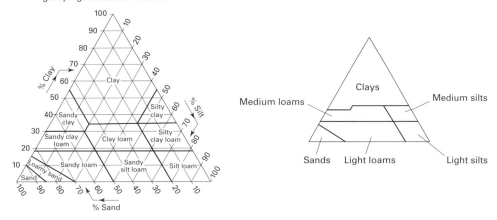

Fig. 2.2 The soil textural triangle used by (a) the USDA and (b) the Soil Survey of England and Wales.

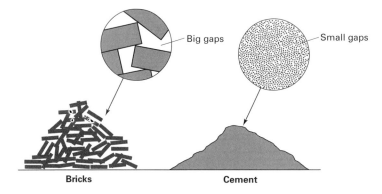

Fig. 2.3 The relationship between particle size and pore size can be illustrated by the analogy of a pile of house bricks and a heap of cement.

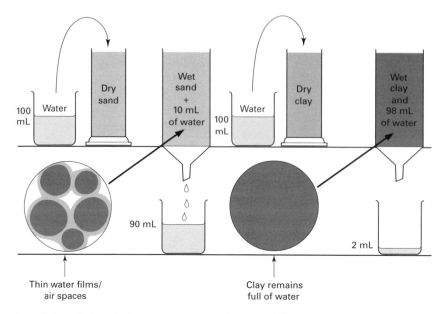

Fig. 2.4 The relationship between pore size and water-holding capacity illustrated by a column of sand and a column of clay.

In sandy soils the gaps between individual grains create a network of large pores, but as the amount of clay increases so does the number of small pores. Figure 2.4 shows one way in which we can relate pore size to soil properties. It shows two hypothetical columns made up of sand and clay (which do you think has the greater number of large pores?). If you were now to pour a beaker of water onto the columns, which column do you think would retain more water? Soils with large pores tend to drain quickly, whereas soils with a high proportion of smaller pores tend to retain water. The number of pores that are water-filled at any one time affects a whole series of soil properties, such as gas exchange with the atmosphere, soil strength and the amount of water held in storage for plants.

Soil texture also has important implications for soil fertility. This is because the proportion of sand, silt and clay along with organic matter largely determines the soil's capacity to store and supply plant nutrients. The nature of these reactions will be discussed in more detail in Chapter 3.

Essential points

♦ The mineral particles in the fine-earth fraction of the soil can be divided on the basis of size. These fractions are called sand, silt and clay.
♦ The proportion of sand, silt and clay in a soil is referred to as its texture.
♦ We can use the soil textural triangle to classify soils into textural groups.
♦ Soil texture affects a whole range of physical and chemical properties.

2   What is meant by the term 'soil structure'?

What is meant by 'soil structure'?

The term 'soil structure' describes the way in which sand, silt and clay particles are bonded together in larger units called 'aggregates'. In order to explain why this is important we need to look again at soil texture. Remember when we used the example of house bricks and builders' cement to illustrate the connection between particle size and pore size? Now you may think that if we were to compare the number of large and small pores in two soils with contrasting textures, one with a clay texture and the other a sandy texture, the sandy soil would have a greater collection of large pores. In some cases this is correct, but in many situations the reverse is true – the number of large pores is often greater in soils that have clay textures.

This may sound confusing at the moment but it is easily explained by the fact that so far we have studied the effects of texture in isolation. At this point we need to look at how particles are arranged together to form aggregates. Soil particles can exist as single loose grains, as an unstructured dense mass or as aggregates of varying sizes. It is the formation of 'soil aggregates' that modifies the basic textural characteristics of a soil, particularly the number of large pores. When we use the term 'soil structure' in its widest sense we are referring to the size, shape and stability of soil aggregates.

The effects of structure can be illustrated by returning to the example we used earlier, when we compared the pore size of a pile of house bricks and builders' cement. Study Fig. 2.5: what happens to the size of gaps between particles if we first mix up the cement with water to form four concrete

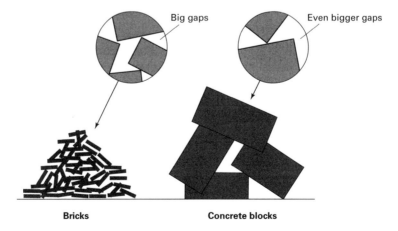

Fig. 2.5 The effects of structure on pore size illustrated by the analogy of house bricks and concrete blocks.

blocks that are much bigger than the house bricks and then randomly stack them together? Which now has the widest gaps between the particles, the bricks or the concrete blocks? It is this tendency to form larger structural units that accounts for the presence of large pores in clay soils.

In soils that do not form aggregates, textural particles can exist either as individual particles, in which case the structure is referred to as being 'single-grained', or, at the other extreme, when the soil exists as one solid block the soil is referred to as 'massive'. In both circumstances, the physical characteristics of the soil will be much more closely related to its texture.

How are soil aggregates formed?

Soil aggregation can be regarded as two processes: aggregate formation and aggregate stabilization. Aggregates are formed when the soil is subjected to shrinking and swelling, plant-root penetration or freezing. All these process tend to break the soil into discrete units. Aggregates are said to be stable when they are able to resist pressures caused by processes such as compaction and sudden wetting. Rapid wetting is a particularly important process in breaking up unstable aggregates because when dry aggregates are suddenly exposed to water, pores near the surface of the aggregate become filled with water, trapping air inside the aggregate; the resulting pressure can sometimes be enough to break the aggregate apart. The disintegration of aggregates by this process is called 'slaking'. Stable aggregates can withstand slaking pressures, whereas unstable aggregates disintegrate into their textural constituents. Figure 2.6 shows how slaking pressures can break up unstable aggregates.

In order to study aggregate formation and stabilization, aggregates are divided into two size groups: microaggregates (<250 μm) and macro-aggregates (>250 μm). Aggregates larger than 1 cm are usually referred to as clods.

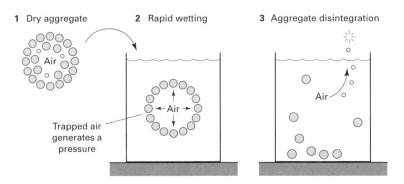

Fig. 2.6 Slaking pressure is generated by air that can become trapped inside aggregates when they are wetted suddenly.

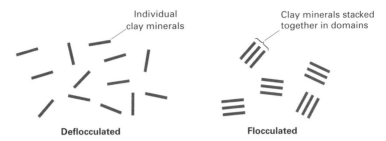

Fig. 2.7 Clay particles can exist in either a deflocculated or a flocculated state when they form small stacks called domains.

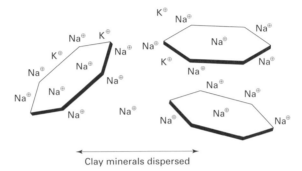

Fig. 2.8 The effect of singly charged ions on clay deflocculation.

How are microaggregates formed and stabilized?

Before microaggregates can form, microscopic clay minerals need to be grouped together in small stacks called 'domains', as shown in Fig. 2.7. When clays are bonded together in this way, they are termed 'flocculated'. Conversely when the clay minerals are dispersed the soil is said to be 'deflocculated'. The first step in the formation of microaggregates is for the clay to be in a flocculated state.

The most important factor influencing flocculation is the presence of ions with more than one charge. Do you remember when we discussed in Chapter 1 how water dissociates to form two ions: H^+ and OH^-? Other important singly charged ions include sodium (Na^+) and potassium (K^+). When clay minerals are covered with singly charged ions they disperse and become deflocculated (Fig. 2.8).

However, not all ions carry one charge. Other ions can have two, three or four charges. For example, calcium (Ca^{2+}) magnesium (Mg^{2+}) and aluminium (Al^{3+}) are three very common ions in soils. Ions with multiple charges are also attracted to the surface of the clay, but unlike singly charged ions, they allow clay minerals to bond together to form domains (Fig. 2.9). Once clay minerals are stacked together to form domains, they can then bond with organic matter to form microaggregates. The process is shown in Fig. 2.10.

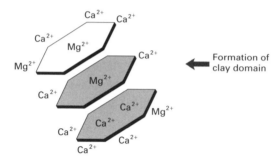

Fig. 2.9 The effect of ions with multiple charges on clay flocculation.

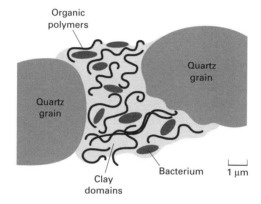

Fig. 2.10 The interaction between clay domains, organic matter, silt and sand particles in the formation of a microaggregate (<250 μm).

The interaction between organic matter and clay domains plays an important role in stabilizing microaggregates. Organic matter associated with microaggregates is not easily altered by changes in land management. This is because once the organic matter has formed a close association with clay minerals, it tends to become more resistant to microbial degradation. It is for this reason that the organic matter associated with microaggregates is referred to as 'persistent'.

How are macroaggregates formed and stabilized?

Once microaggregates have formed they can then coalesce to form macro-aggregates. The idea that large aggregates are simply collections of micro-aggregates, bound together with organic matter, is called the 'aggregate hierarchy model'. It is shown in Fig. 2.11. In most circumstances they do not coalesce into one massive block, but into several larger aggregates. This is because environmental factors such as drying, freezing and plant-root penetration tend to exploit planes of weakness. This tends to create a range of discrete macroaggregates rather than one massive, unstructured mass of soil.

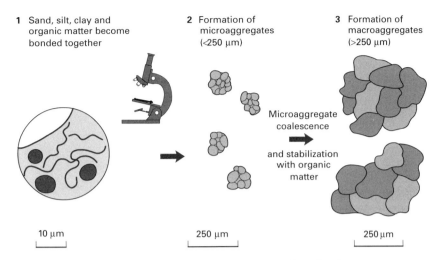

1 Sand, silt, clay and organic matter become bonded together

2 Formation of microaggregates (<250 μm)

3 Formation of macroaggregates (>250 μm)

Microaggregate coalescence

and stabilization with organic matter

10 μm

250 μm

250 μm

Fig. 2.11 The formation of macroaggregates (>250 μm) is explained in the aggregate hierarchy model.

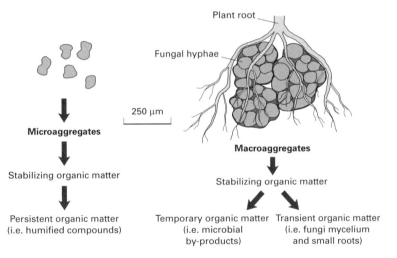

Plant root

Fungal hyphae

250 μm

Microaggregates

Stabilizing organic matter

Persistent organic matter (i.e. humified compounds)

Macroaggregates

Stabilizing organic matter

Temporary organic matter (i.e. microbial by-products)

Transient organic matter (i.e. fungi mycelium and small roots)

Fig. 2.12 The role of persistent, temporary and transient organic stabilizing agents in the stabilization of micro- and macroaggregates.

In soils that have low concentrations of clay, macroaggregate stability is highly dependent on organic matter. The type of organic matter associated with macroaggregates is slightly different from the persistent organic material found in microaggregates. Macroaggregate organic matter can be divided into two types, as shown in Fig. 2.12. Firstly, there are those stabilizing agents that are referred to as 'temporary'. These consist of microbial and plant by-products, the most important of which are the 'polysaccharide gums' that are simply long chains of sugar molecules. Secondly, there are 'transient' stabilizing agents, which include fine plant roots and fungal

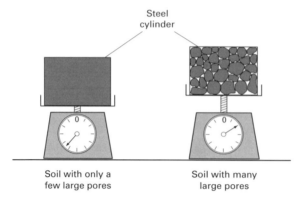

Fig. 2.13 The structure of the soil can be described by measuring its bulk density. This is simply a measurement of the mass of particles in a given volume. We can take a bulk density measurement by pushing a steel cylinder (with a known volume) into the ground; the soil can then be removed, dried and weighed and its bulk density calculated.

hyphae. Both temporary and transient stabilizing compounds are vulnerable to microbial attack so need to be replenished continuously through inputs of fresh soil organic matter.

The arrangement of aggregates and pores is a dynamic, not a static, quality. Shrinking and swelling, plant-root extension, tillage and freeze–thaw effects are continually altering the arrangement of pores and aggregates. This is coupled with the slow mineralization of temporary and transient organic binding agents, which leads to the slow disintegration of macroaggregates. However, as old aggregates fall apart new ones are created and stabilized.

How can we describe soil structure?

We now understand what soil structure is, but how can we describe it? Look at Fig. 2.13: one of the soil cores has only a few large pores whereas the other has many. Given that both soils occupy the same volume, which one do you think weighs more?

The answer is that the soil with the greater amount of soil particles and the least pore space in a given volume weighs more. This means that we can use the ratio of weight to volume to calculate the amount of pore space in a soil. Soil scientists refer to this property as the soil's 'bulk density'. Bulk density is easily measured by pushing an open-ended steel cylinder into the ground. The cylinder can be any size as long as its volume is known. Once the cylinder is flush with the soil surface it can be dug up and its contents removed, dried and weighed. Bulk density can then be calculated with the following equation:

$$\text{Bulk density (g/cm}^3) = \frac{\text{Mass of soil (g)}}{\text{Volume of cylinder (cm}^3)}. \tag{1}$$

Soils with higher bulk densities have less pore space. The formation of stable macroaggregates with many large pores is the reason why soils with high clay concentrations often have lower bulk densities than sandier soils. We can illustrate how bulk density is measured by using a hypothetical example. A steel cylinder with a volume of 177.6 cm^3 is hammered into the soil so that its top is level with the soil surface. The contents of the cylinder are then removed and dried in an oven at 105°C before being weighed. In this case we find that the weight of soil is 248.7 g. We can calculate the bulk density of the soil as follows:

$$\left[\frac{248.7 \text{ g (mass of soil)}}{177.6 \text{ cm}^3 \text{ (volume)}} \right] = 1.4 \text{ g/cm}^3. \tag{2}$$

In this case it means that there are 1.4 g of particles in every cm^3 of soil.

The porosity of the soil can be calculated simply by using the bulk density measurement and the average density of soil particles, which is 2.7 g/cm^3:

$$\text{Porosity } \% = \left[1 - \left(\frac{\text{Bulk density}}{2.7} \right) \right] \times 100. \tag{3}$$

Using the bulk density of 1.4 g/cm^3 value we calculated above, we can calculate the porosity of the soil:

$$\text{Porosity } \% = \left[1 - \left(\frac{1.4}{2.7} \right) \right] \times 100 = 48\%. \tag{4}$$

In this case 48% porosity means that 52% of the soil volume is occupied by soil particles. Values of 35–50% porosity are typical for a sandy soil, and 40–60% for clay-textured soils.

The ideal situation for plant growth is to have a combination of small (micropores), medium (mesopores) and large (macropores) pores; this is because they are important for different reasons: mesopores and micropores for water storage and macropores for gas exchange.

Essential points

♦ Textural particles can be bonded together in larger structural units called aggregates.
♦ Soil aggregates are formed by drying, root penetration and freezing and then stabilized by clay mineral attraction and organic matter.
♦ Aggregates can be divided into microaggregates (<250 μm) and macroaggregates (>250 μm).
♦ Microaggregates are stabilized by persistent organic matter whereas macroaggregates are stabilized by transient and temporary organic matter.
♦ The combined effects of soil texture and structure determine the range of pore sizes within the soil. This can be described by measuring the soil's bulk density.

3   How does water behave in the soil?

What is soil water?

What happens if you immerse a sponge in a bucket of water and then hold it up in the air? After an immediate stream of water, the water loss slows to a few drips: if you squeeze it again another stream of water pours out. You can carry on applying pressure until the sponge is wrung out. The sponge is not dry, however; it's just that the remaining water is held so tightly that you cannot generate enough pressure using your fingers alone to force the water out of the sponge. So, how can we relate a wet sponge to soil water?

We now know that soils contain a whole series of pores that vary in size. Pore size is important because it determines water retention. Large pores hold water loosely while small pores hold water very tightly. Soils may be saturated immediately following prolonged rainfall, but as soon as the rainfall ceases large pores begin to drain. The soil will still remain wet to the touch because water will be retained in medium- and small-sized pores (just like the sponge before it was squeezed). Pore size therefore controls the tightness with which water is held. This tightness is referred to as 'tension' and is a measure of how much suction the soil pore exerts on the water. Figure 2.14 shows the gradual decrease in the size of pores filled with water as a loamy textured soil drains over 7 days.

How does pore size affect water retention?

At a very simple level, we can divide soil pores into three size classes depending on their role in the movement and storage of water. The largest, macropores, or 'transmission pores', allow the rapid movement of water and gas through the soil. Water in these large pores usually drains within 48 hours. Mesopores or 'storage pores' hold water against the force of gravity; they are particularly important in storing water for plants. Finally, micropores or 'residual pores' hold water very tightly. Although this water can still play a part in chemical reactions it is too tightly held for plant use. Table 2.2 shows the size range of each pore-size class.

Soils with differing textures and structures will have varying quantities of macro-, meso- and micropores. A combination of how well these pores are connected to each other and the pore-size range will determine the moisture characteristics of the soil. In order to understand how pore size, water movement and water storage are all related, scientists have studied how water behaves in saturated soils and then how its behaviour is modified under unsaturated conditions.

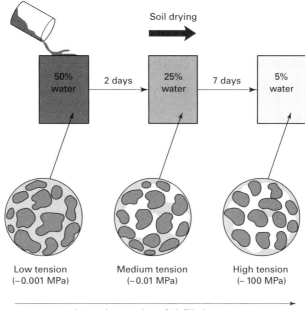

Fig. 2.14 The relationship between water-filled pore size and water tension can be illustrated by considering what happens when a soil begins to dry from a saturated state. Here we have chosen a soil with a loam texture. However, these values will not apply in soils that have different textures; this is because their pore-size distribution will be different.

Table 2.2 Pore-size classification.

Class	Pore size	Function
Transmission	Macropores (>50 μm)	Aeration and plant-root extension
Storage	Mesopores (0.5–50 μm)	Storage of plant-available water
Residual	Micropores (<0.5 μm)	Tightly held water important for some soil structural properties

What forces are acting upon the soil water in a saturated soil?

When a soil is saturated, all the soil pores, from the largest to the smallest, are filled with water. Under these conditions soil water tension is zero. This situation usually occurs after prolonged heavy rainfall or under waterlogged conditions.

As the water infiltrates the soil the leading edge of water (called the 'wetting front') moves down the profile. The speed at which water percolates through the soil is determined by the soil's 'hydraulic conductivity'. Basically, this measurement refers to how easily the water can move through the soil profile. This will vary between different soil types, because water flow will depend upon two opposing properties: the pressure generated by the water in the soil and the resistance to water flow caused by the soil particles. Hydrologists sometimes refer to the pressure component as 'total head' because it is made up of several factors, such as elevation and the pressure generated by the weight of the water column overlying the wetting front. The resistance to water flow in the soil is determined by three factors: how well the pores are connected together; the pore-size distribution; and the water content of the soil. In soils that have many small air-filled pores, the movement of water can be very slow; in other soils that have higher amounts of large water-filled pores, water movement can be rapid.

The factors that determine hydraulic conductivity were first described mathematically in 1856 by a French civil engineer called Henri Darcy. He secured himself a place in history after working on the problem of predicting water movement through sand filters for the city of Dijon. The equation he developed (referred to as 'Darcy's law') can be stated as 'discharge is directly proportional to the loss in head'. This relationship is used extensively to describe water movement in a number of media, including ground water.

Once the largest transmission pores have emptied of water, usually within 48 h, the soil is said to be at 'field capacity'. The water that is lost over this period is referred to as 'gravitational water'. Once field capacity has been reached the remaining water is held in the soil under tension in mesopores and micropores.

What forces are acting upon the soil water in an unsaturated soil?

As the soil drains, the forces acting upon the soil water change. As drying continues, large pores empty and become air-filled as water is confined to progressively smaller pores. Eventually the water is held very tightly in micropores as 'residual' or 'hygroscopic' water.

We can illustrate the connection between pore size and water tension simply by pouring water into a glass bowl: under normal atmospheric conditions it has zero tension, just like water in a saturated soil. If we wanted to, we could easily suck the water out of the bowl using a straw. If we were now to pour a bag of dry clay into the bowl, so that a thick mud pie is formed, we would find that, by combining with the clay, the water is now held under tension. It now takes considerably more energy to extract the water from the clay (just think about how much water you could suck up from the bowl now).

Essential points

♦ Water movement in saturated soil is determined by the total head and hydraulic conductivity of the soil. As the soil drains from a saturated state, water is lost from the largest pores first: this is referred to as gravitational water.

♦ Under non-saturated conditions water will be held under tension in mesopores and micropores.

♦ Pore size will affect the tension by which soil water is held. Water held in mesopores is referred to as storage water and water held in micropores is termed residual water.

4   Why are the physical characteristics of the soil so important?

Why is soil structure important?

When soil aggregates are unstable they are unable to withstand slaking pressures or compaction. Both processes can lead to poor water infiltration and gas exchange, both of which are important for plants.

The stability of macroaggregates can only be maintained if there is a continuous replenishment of organic matter to replace the temporary and transient binding agents that are constantly being degraded by soil microbial organisms. If the replenishment of these compounds is interrupted, macroaggregate stability declines. Figure 2.15 shows one example where aggregate stability has been left to decline in experimental plots that have not received organic matter.

Reductions in macroaggregate stability and the subsequent loss in macropores can reduce the infiltration rate of water and the exchange of gases with the atmosphere. If organic matter losses continue over a long time, microaggregate stability will also eventually decline, as persistent compounds are slowly mineralized. Declines in aggregate stability are commonly found when grassland sites are cultivated for the first time. The effects of cultivation are twofold. Firstly, in many agricultural systems fewer plant residues

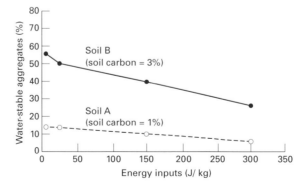

Fig. 2.15 The relationship between soil organic matter and aggregate stability. (From C. W. Watts and A. R. Dexter (1997) The influence of organic matter in reducing the destruction of soil by simulated tillage. *Soil & Tillage Research* **42**; with permission from Elsevier Science.)

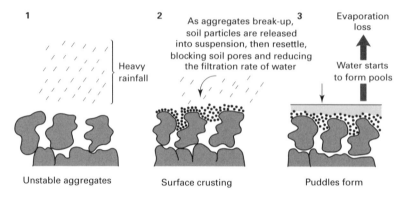

Fig. 2.16 Soil crusting: heavy rainfall can break apart unstable aggregates, releasing mineral particles that then block pores. This can slow down the infiltration rate and lead to high evaporation losses.

are returned to the soil. This leads to a decline in the amount of soil organic matter. Secondly, cultivation tends to break apart aggregates, exposing the temporary and transient organic matter to microbial attack.

Aggregate instability is a particularly serious problem in soils that have high proportions of sand and silt. Aggregates in these soils can slake even under the mildest wetting pressures. As aggregates break open, sand, silt and clay particles are released and washed into soil pores, preventing further water infiltration: this process is called 'soil crusting' (see Fig. 2.16). Crusting effectively seals the soil surface so that instead of infiltrating the soil, rain water collects in puddles where it is then evaporated. In severe cases run-off is concentrated into small channels called 'rills'. These can deepen and lead to serious soil erosion.

Why is soil water tension important?

Plants can only extract water at certain tensions; if the water is held too tightly it is unavailable for plant uptake. Remember the example we used earlier in this chapter when we poured a bag of clay into a bowl of water before trying to suck water from the mud pie using a straw. A similar situation occurs in soil because plants can only exert a limited amount of energy to extract water. Since textural and structural differences will affect the pore-size distribution, the amount of plant-available water will differ depending on the relative amount of micro-, meso- and macropores. This is illustrated in Fig. 2.17.

The easiest way to describe this relationship is if we plot water content against water tension; this is referred to as a 'moisture release curve'. Tension is usually measured in units such as pascals, bars or centimetres. Table 2.3 shows the relationship between the three types of unit.

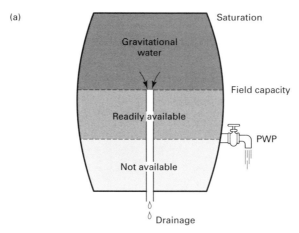

The barrel represents the soil moisture characteristics of a clay soil

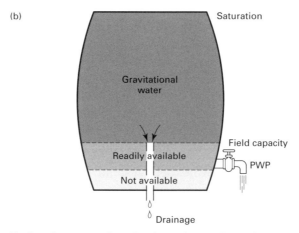

The barrel represents the soil moisture characteristics of a sandy soil

Fig. 2.17 The soil moisture release curve can be illustrated by the analogy of a barrel. Here we have shown two barrels, (a) from a clay, and the other (b) from a sandy-textured soil. PWP, or permanent wilting point – the point at which the plant can no longer extract water from the soil. (From L. D. Doneen and D. W. Westcot (1984) Irrigation practice and water management. *Irrigation and Drainage Paper* **1**. FAO, Rome.)

Table 2.3 The relationship between pressure measurements.

Height of water column (cm)	Water potential (bars)	Water potential (MPa)
0	0	0
10	−0.01	−0.001
102	−0.3	−0.03
1 020	−1.0	−0.1
15 300	−15	−1.5

Source: N. C. Brady (1990) *The Nature and Properties of Soils*. Macmillan, New York.

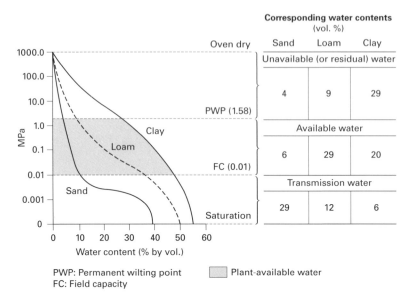

Fig. 2.18 The moisture release curve can be used to illustrate how texture influences the tension at which water is held. (Adapted from J. R. Landon (1991) *Booker Tropical Soil Manual: A Handbook for Soil Survey and Agricultural Land Evaluation in the Tropics and Subtropics.* Longman Scientific and Technical, Harlow.)

The moisture release curves in Fig. 2.18 are for three soils with differing textures. The amount of plant-available water lies between two points; at the wetter end of the curve, water held in storage pores when the soil has come to field capacity is usually held at tensions around 0.01 MPa. The drier end of the curve is called the wilting point. Here, the water is held under a great deal of tension, typically 1.5 MPa (which is the same as the tension you get when drawing water up a 150 metre straw!); beyond this point the water tension becomes so high that the water cannot be used by the plant. Compare the amount of plant-available water in the clay-textured soil with that of the loam and sand. See how the sandy soil has the smallest amount of stored water that the plant can use.

Ideally there should be a balance between large macropores for gas exchange and smaller pores for water storage. Soils with loamy textures (equal combinations of sand, silt and clay) provide some of the best agricultural soils because they provide enough water storage for plant growth and sufficient large pores for gas exchange with the atmosphere.

Essential points

♦ Macroaggregate stability is essential to prevent the collapse of macropores and mesopores.
♦ Both microaggregate and macroaggregate stability is dependent on the continuous supply of organic matter.

♦ Cultivation can lead to declines in the amount of organic matter returned to the soil. This can cause reductions in the stability of aggregates.

♦ Soils that have low aggregate stability tend to suffer problems such as surface crusting and compaction.

♦ The availability of water for plant growth is determined by textural and structural properties of the soil.

♦ The amount of water a soil can store can be described in terms of plant growth using three classifications: transmission, storage and residual water.

♦ The amount of plant-available water is the amount of water that lies between field capacity and wilting point. This can be shown graphically using a moisture release curve.

♦ The optimum condition for plant growth is for the soil to have a combination of micropores and mesopores for water storage and macropores for gas exchange with the atmosphere.

Chapter Summary

The proportion of sand, silt and clay in a soil is referred to as its soil texture. The textural composition of the soil affects its physical and chemical properties. However, some of the physical properties associated with texture can be modified because soil particles tend to bond together into larger units called aggregates. Once aggregates are formed they are then stabilized by organic matter. Stabilization allows aggregates to survive moderate compaction and slaking pressures. In order to study the stabilization of aggregates, scientists have divided them into two size groups: microaggregates (<250 μm) and macroaggregates (>250 μm). The pore-size distribution in the soil is dependent on its texture and structure. Pore size is important because it determines how tightly soil water is held. Small pores exert high tension and large ones low tension. We can express the relationship between water content and tension using a moisture release curve. The two most important points on the moisture release curve are field capacity and wilting point. The water content between these two points is referred to as plant-available water. Plant-available water is a measurement of how much water a soil can store for plant growth under non-saturated conditions.

3   Soil Surfaces, Acidity and Nutrients

Introduction

Unlike animals, plants are immobile organisms. Being stationary they cannot move to seek food or escape harsh conditions; they must therefore rely on the soil to provide them with nutrients and protection. As well as offering benefits, soils can also act as a store of acidity and pollutants. The potential for a soil to store chemicals depends upon its texture (specifically the amount of clay it contains) and its organic matter content. The ability of clay minerals to store chemicals is determined by two important properties: surface area and surface chemistry. Some scientists have speculated that clay minerals may have played an important role in the development of 'life' on earth. This theory is based on the way clay minerals can concentrate and protect chemicals from degradation.

This chapter will look at the ways by which clay minerals and organic matter store and release chemicals. It will do this by following two main lines of enquiry:

1   **How are chemicals stored and released from soil?**
   What are colloids and why are they important?
   Clay mineral structure
   Soil humus composition
   How does the electrostatic charge differ between colloidal particles?
   What is the source of permanent charge?
   What is the source of variable charge?
   How do soil colloids affect the adsorption and release of chemicals?

2   **What determines the availability of acids and nutrients?**
   What is soil acidity?
   Problems associated with acid soils
   What are soil nutrients?
   How is the availability of nitrogen, potassium and phosphorus affected by soil colloids?
   How can we describe a soil's ability to store and release nutrients?

1   How are chemicals stored and released from soil?

The chemicals soil scientists are really interested in are those that have a beneficial effect on plant growth (such as nutrients) and those that are toxic to plants and animals. The reactions that affect the availability of beneficial and harmful chemicals are similar. This is because soil chemistry is largely governed by the soil's 'colloidal' material.

What are colloids and why are they important?

When soil scientists talk about 'colloidal particles' they are usually referring to clay minerals and soil organic matter or, more specifically, the humus fraction. In strictly chemical terms, colloids are particles that are larger than individual molecules but small enough to remain microscopic. When colloidal particles are mixed with water they remain in suspension in a dispersed state. Their unique behaviour is due to their enormous surface area to volume ratio or 'specific surface area' and their electrostatic properties. Figure 3.1 shows how we can divide soil particles into those with and those without colloidal properties.

We can illustrate the importance of surface area by taking a 1 m^3 cube that has a total surface area of 6 m^2. If we were to subdivide the cube into microscopic cubes, each with sides which are one-millionth of a meter long, we find that the total surface area has increased to 6 000 000 or 6×10^6 m^2. Similarly, it has been estimated that 1 cm^3 of charcoal has a surface area of 10 000 000 cm^2, which is why it is such an effective filtering

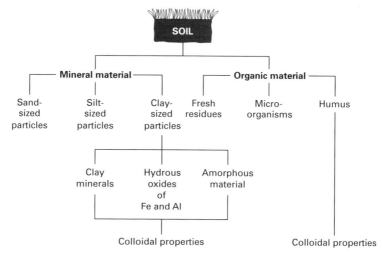

Fig. 3.1 Soil is made up of a mineral and an organic matter fraction. Both fractions contain colloidal material. It is these colloidal materials that have a great influence on the chemical properties of the soil.

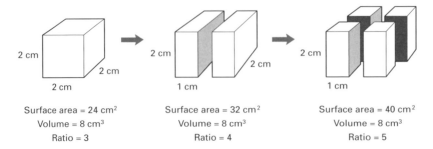

Fig. 3.2 The relationship between particle size and surface area can be demonstrated if we repeatedly divide a cube. Note how the volume stays the same as the surface area increases.

agent. The relationship between particle size and surface area is shown in Fig. 3.2, where a cube with an initial surface area of 24 cm^2 is repeatedly divided.

In addition to their surface area, many of the soil colloids also carry an electrostatic charge. In the last chapter we divided soil into the following fractions: sand, silt, clay and SOM (soil organic matter, including the humus fraction). If we do this again, but instead of looking just at size differences we also consider whether the particle is charged or not, we find that of the four fractions only the clay-sized fraction, including clay minerals, hydrous oxides of iron and aluminium, and soil humus have an electrostatic charge. This gives them the ability to hold on to certain chemicals, so that they can store nutrients, acidity and toxins. It is the presence of these colloidal particles that largely controls the storage and release of chemicals in the soil.

In order to appreciate their importance we must first consider why they possess an electrostatic charge. We will then look at how this charge governs the behaviour of chemicals in the soil.

Clay mineral structure

Do you remember when we introduced soil texture and the clay-sized fraction in Chapter 1? At that stage we were content with a basic description; now we need to look in a bit more detail at how the clay-sized particles differ from each other and how these differences affect the type of charge they possess.

Soil mineral colloids in the clay-size fraction consist of clay minerals, hydrous oxides of iron and aluminium (metals which have combined with oxygen and water) and minerals with no clearly definable structure, which are usually referred to as 'amorphous'. We will concentrate on the clay minerals, because they are the most important mineral colloids in the majority of the world's soils.

Silicate clays can be divided into several groups, but we shall only look at the two most important ones, known as 1 : 1 and 2 : 1 clay minerals. The

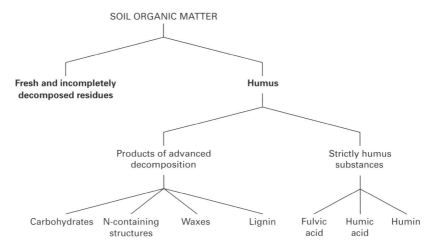

Fig. 3.3 Soil organic matter can be fractionated into the following components: fresh and incompletely decomposed residues, and humus. Humus comprises two fractions: products of advanced decomposition and strictly humic substances. (Adapted from M. M. Kononova (1966) *Soil Organic Matter*. Pergamon, Oxford.)

difference between them is simply that 1 : 1 clays have one layer of silica to one layer of aluminium hydroxide (like butter on a slice of bread), whereas 2 : 1 minerals have two layers of silica to one layer of aluminium hydroxide (like a sandwich). Within this broad division there are several different types of clay mineral, examples including montmorillonite (which is part of the smectite group of clay minerals) and illite (which is sometimes also called hydrous mica) which are both 2 : 1 clays but with slightly different structures. This causes them to have different physical and chemical properties, the main one being that montmorillonite clays, unlike illite, can shrink and swell depending on moisture content. Furthermore, the division of clay minerals into those with 1 : 1 and 2 : 1 structures is useful because they also have different charge characteristics. This influences the way chemicals are stored and released from the soil solution.

Soil humus composition

We can divide soil organic matter into several groups of compounds, ranging from fresh and partially decomposed residues to substances classed as 'strictly humus substances'. Figure 3.3 summarizes these divisions.

Soil humus consists of two fractions; 'products of advanced decomposition' and 'strictly humic substances'. Both fractions are composed of a whole range of chemical compounds, such as hydroxyl (OH^-), carboxyl (COO^-), phenol (C_6H_5OH) and benzene (C_6H_6) groups. Unlike the structure of clay minerals, the chemical composition of humus has proved to be extremely difficult to unravel. This is partly because its composition varies with changes in soil conditions. Despite these problems, soil chemists have

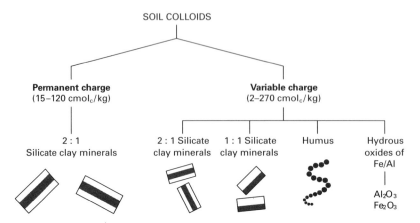

Fig. 3.4 If we divide soil colloids on the basis of their charge characteristics, we find that 2 : 1 clay minerals have both a permanent charge and a variable charge, whereas 1 : 1 clay minerals, humus and oxides of iron and aluminium have only a variable charge.

managed to develop ways of dividing the strictly humic substances into several groups. These divisions are based on the degree of polymerization (a process by which molecules are joined together) in the compound, and the main chemical structure of carbon atoms or – more specifically – whether the carbon atoms are joined together in long chains or rings.

For example, in the case of 'fulvic acids' the degree of polymerization is low, and the carbon atoms are mainly joined together to form chain structures. This is unlike the 'humic acid' fraction that contains high concentrations of highly polymerized structures in which carbon atoms are joined together in ring structures.

How does the electrostatic charge differ between colloidal particles?

Not only do clay minerals and humus molecules differ in their size, shape and composition, but they also differ in terms of their electrostatic properties (Fig. 3.4). The electrostatic charge associated with soil colloids can be either 'permanent' or 'variable'. If we divide the soil's colloidal particles on the basis of whether the main charge is permanent or variable, we find that 2 : 1 clays possess both a 'permanent' and a 'variable' charge, whereas 1 : 1 clay minerals, hydrous oxides of iron and aluminium and humus have mainly a 'variable' charge. The charge characteristics of humus, and 2 : 1 and 1 : 1 clay minerals are summarized in Table 3.1.

The main properties of the permanent charge can be summarized as follows:

♦ it is not affected by soil acidity;
♦ it is mainly a property of the 2 : 1 clays;
♦ it is nearly always negative.

Table 3.1 Soil colloids and their charge characteristics.

Colloid	Permanent (%)	Variable(%)	Total (cmol/kg)
Organic matter	10	90	200
Montmorillonite (2 : 1)	95	5	100
Kaolinite (1 : 1)	5	95	8

Source: N. C. Brady (1990) *The Nature and Properties of Soils*. Macmillan, New York.

All soil colloids have variable charge. As the name suggests, variable charge can be positive, negative or zero, depending upon the acidity of the soil solution. Characteristics of variable charge can be summarized as follows:
♦ the 2 : 1 and 1 : 1 clays, hydrous oxides of iron and aluminium and soil humus all have a variable charge;
♦ variable charge can be positive, negative or zero;
♦ variable charge is strongly influenced by soil pH.

What is the source of permanent charge?

Both 2 : 1 and 1 : 1 clay minerals are made up of layers of silica and aluminium hydroxide. The silica layer consists of a series of silicon and oxygen atoms, in the ratio of 1 : 4, forming small pyramid-shaped structures called 'silica tetrahedra'. When these consist of pure silicon and oxygen they form the mineral 'quartz', which is highly resistant to chemical weathering (most of the sand grains found in soils are composed of quartz). However, clay minerals form under a range of conditions and often have elements other than oxygen and silicon. Sometimes atoms that are similar in size to silicon combine with oxygen to form part of the silica sheet. Although similar in size, their chemistry is different from that of silicon. A similar situation can also occur in the aluminium layer, where magnesium and iron can substitute for aluminium. This process is called 'isomorphic substitution' and is summarized in Fig. 3.5.

Isomorphic substitution often causes a charge imbalance within 2 : 1 clay minerals, which results in a permanent negative charge. Sometimes the substitution occurs mainly in the silica sheet and other times in the aluminium hydroxide sheet, depending on the clay mineral. This distinction is important because it influences the properties of the 2 : 1 mineral. For example, in the case of illite much of the isomorphic substitution occurs in the silica sheet, which creates a relatively localized charge that is neutralized by potassium ions. These link adjacent clay minerals tightly together. In montmorillonite, it is the aluminium hydroxide sheet that is heavily substituted. This gives rise to a more delocalized charge that allows water to penetrate between individual minerals, allowing them to shrink and swell depending on the water

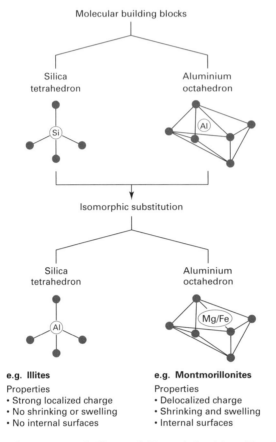

Fig. 3.5 Clay minerals are composed of layers of silica and aluminium. The silica and aluminium layers are constructed differently: the silica layers are composed of interlinked tetrahedra, whereas the aluminium layer is constructed of interlinked octahedra. We can relate changes in the crystal structure to differences in the physical and chemical properties of clay minerals.

content. This allows montmorillonite to adsorb cations on its internal, as well as external surfaces. This greatly increases its ability to store chemicals when compared with other clay minerals that only possess external surfaces. The role of internal and external surfaces in cation adsorption is shown in Fig. 3.6.

Isomorphic substitution is not a significant feature of 1 : 1 clays, which is why they have only small amounts of permanent charge.

What is the source of variable charge?

The origins of variable charge have nothing to do with isomorphic substitution but result from processes we have already covered in Chapter 1. Do you remember when we discussed the dissociation of water molecules into two charged particles, called a hydroxyl and a hydrogen ion? The process is shown in Fig. 3.7.

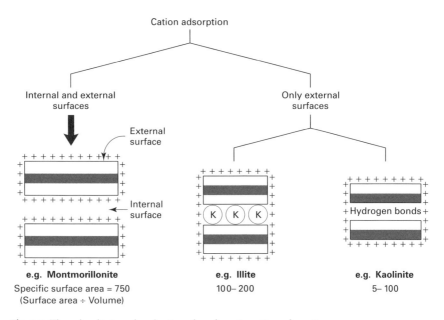

Fig. 3.6 The role of internal and external surfaces in cation adsorption.

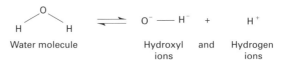

Fig. 3.7 Water dissociation. It should be noted that this is a simplified version of the reaction because in reality H^+ ions in water exist as H_3O^+.

The hydroxyl group, which consists of one hydrogen joined to one oxygen atom, can occur in isolation, as it is often found in soil solution, or as part of a larger molecule (usually abbreviated to R-OH (where 'R' simply symbolizes the rest of the molecule). When it occurs as part of a larger structure it can dissociate (break up) to leave a negatively charged O^- group attached to the larger structure (R-O^-) and a positively charged H^+ ion in solution. In addition to forming negatively charged sites, when there are high concentrations of H^+ ions in solution (as in acid soils), H^+ ions can combine with R-OH^- groups to form R-OH_2^+; this process is referred to 'protonation'. It is not only hydroxyl groups that can dissociate or become protonated depending on the H^+ concentration of the soil solution – often other chemical groups such as carboxyl and phenol groups can also undergo similar reactions (Fig. 3.8).

All soil colloids possess chemical groups that can undergo dissociation and protonation. In the clay minerals they are often associated along the edges of the crystal, whereas in organic matter the variable charge is the result of the chemical groups it contains, many of which can undergo dissociation and protonation depending on the soil solution. This creates

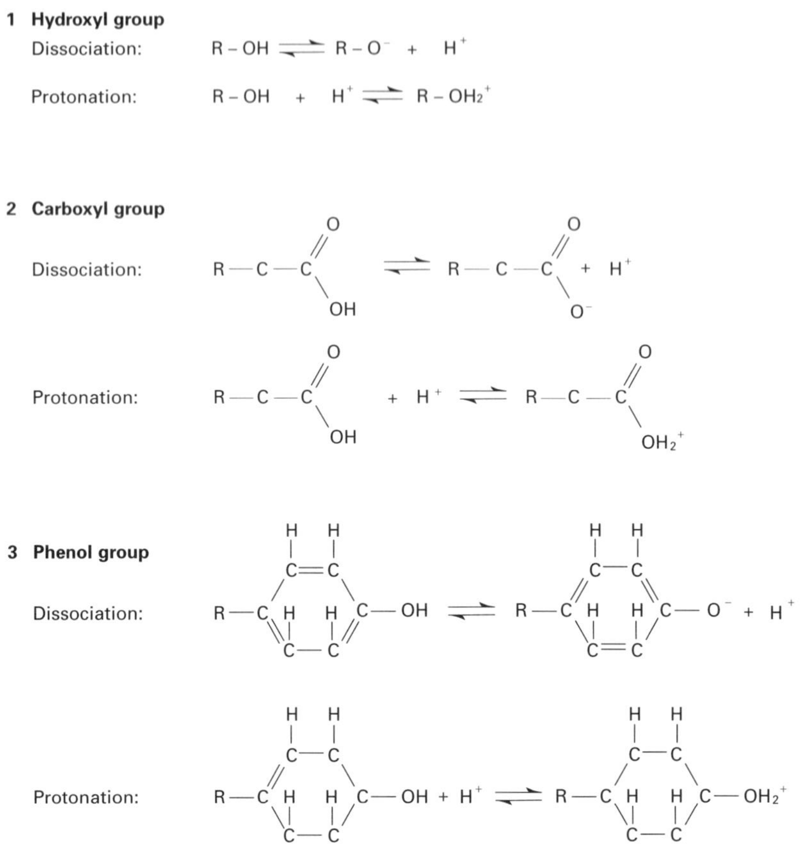

1 Hydroxyl group

Dissociation: $R-OH \rightleftharpoons R-O^- + H^+$

Protonation: $R-OH + H^+ \rightleftharpoons R-OH_2^+$

2 Carboxyl group

Dissociation: $R-C-C\big(\!\!\overset{O}{\underset{OH}{}} \rightleftharpoons R-C-C\big(\!\!\overset{O}{\underset{O^-}{}} + H^+$

Protonation: $R-C-C\big(\!\!\overset{O}{\underset{OH}{}} + H^+ \rightleftharpoons R-C-C\big(\!\!\overset{O}{\underset{OH_2^+}{}}$

3 Phenol group

Dissociation: $R-C_6H_4-C-OH \rightleftharpoons R-C_6H_4-C-O^- + H^+$

Protonation: $R-C_6H_4-C-OH + H^+ \rightleftharpoons R-C_6H_4-C-OH_2^+$

Fig. 3.8 Some other important reactions involving the dissociation and protonation of chemical groups.

areas of electrostatic charge. Figure 3.9 shows the structure of a highly simplified section of soil organic matter.

How do soil colloids affect the adsorption and release of chemicals?

We now understand what soil colloids are, how the type of electrostatic charge differs between colloids and the source of this charge. Before leaving this section we must consider how the colloidal surface interacts with soil solution.

At the colloid's surface, ions are held very tightly. However, successive layers of ions, held at increasing distances from the charged surface, are less tightly held. For most soils, which are not too acidic or alkaline, the charge associated with both permanent and variable sites will be negative. This attracts cations to collect in a tightly held layer at the surface of the colloid and a less tightly held 'cloud' of cations further away. Figure 3.10 shows how we can divide this ionic cloud into two sections.

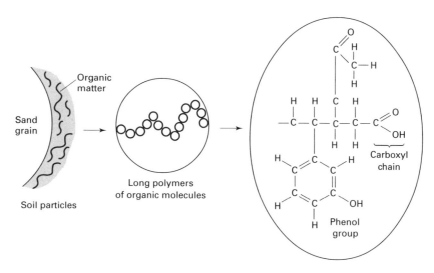

Fig. 3.9 Organic matter possesses a variable electrostatic charge because it contains a whole range of chemicals that can undergo dissociation and protonation, depending on the soil solution.

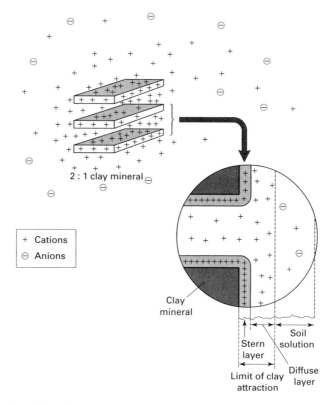

Fig. 3.10 We can show the surface chemistry of clay minerals as a series of layers containing cations and anions.

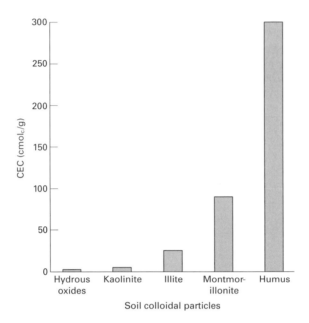

Fig. 3.11 The CEC (cation exchange capacity) of several soil colloidal particles.

The innermost layer is referred to as the 'Stern layer'. The outer, less tightly held layer is referred to as the 'diffuse layer'. Cations can be exchanged between each layer and the soil solution as the concentration of cations in soil solution changes. The relative strength by which cations are held is dependent on the charge characteristics of the ion. The strength of attraction in some common ions is shown below:

$$Al^{3+} > Ca^{2+} > Mg^{2+} > K^+ = NH_4^+ > Na^+$$

$$\longrightarrow \qquad\qquad\qquad (5)$$

Decreasing strength of attraction

The ability of the soil to adsorb cations is referred to as its 'cation exchange capacity' or CEC. This property is of vital importance when attempting to measure a soil's ability to supply nutrients such as K^+, Ca^{2+} and Mg^{2+}. Soils that have high CECs tend to have high concentrations of colloidal particles (particularly 2 : 1 clay minerals and organic matter) and a neutral to high pH (>7). For example, a clay-rich pasture soil will have a far higher CEC than an acidic sandy-textured soil. The CEC of a soil is therefore a measurement of the total amount of permanent and variable exchange sites a soil has. Figure 3.11 shows the CEC of several soil constituents.

Anions, such as NO_3^-, are not attracted to colloidal surfaces because like charges repel. It is for this reason that many anions are susceptible to leaching. However, certain exceptions do occur. One of these is under high levels of protonation, when sites with variable charge become positively charged. When this happens anions can be attracted directly to the colloidal surface. The other exception occurs when anions such as PO_4^{2-} bond with other simple soil chemicals that modify its basic anion characteristics (ligand formation). This allows them to become adsorbed on the clay surface. For a more detailed explanation of these mechanisms, refer to one of the specialist texts in the reading list.

Essential points

♦ Soil colloids include the clay minerals, hydrous oxides of iron and aluminium, amorphous material and humus. Soil colloidal particles have an enormous surface area and an electrostatic charge.

♦ The charge associated with soil colloids can be permanent, resulting from isomorphic replacement within the mineral structure, or variable, resulting from the dissociation and protonation of chemical groups along the edges of clay minerals and in organic matter. In most situations both the permanent and variable charge will be negative.

♦ 2 : 1 clay minerals have both a permanent and variable charge. In 1 : 1 clays, organic matter and hydrous oxides the charge is mainly variable.

♦ Cations are attracted to the negatively charged soil colloids, forming a layer of positively charged particles.

♦ The cation cloud can be divided into two sections: first, there are the cations that are held tightly at the colloid's surface (these form the Stern layer). Second, there are cations held further away from the colloid's surface which form a diffuse layer. Cations can exchange between these sections and the soil solution through ion exchange.

♦ The ability of a soil to store nutrients is related to both the amount and the type of colloidal material present. The ability of a soil to store nutrients can be expressed in terms of its cation exchange capacity or CEC. Soils with large amounts of clay and organic matter have high CECs whereas soils with sandy textures and low concentrations of organic matter have low CECs.

♦ As well as external surfaces some clay minerals, such as montmorillonite, have internal surfaces. This increases their overall surface area.

♦ Many anions are susceptible to leaching because, unlike cations, they are not attracted and adsorbed directly by colloidal surfaces. However, exceptions do occur, as some anions can have their basic chemical properties altered by other soil chemicals.

2   What determines the availability of acids and nutrients?

The simplest answer to this question is that soil colloids have the ability to act as a chemical store because they 'buffer' chemical changes by soaking up and then releasing ions from the soil solution. We will start by looking at how soil colloids affect soil acidity.

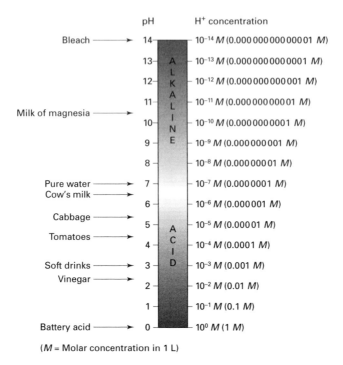

pH   H⁺ concentration

Bleach ⟶ 14 — 10⁻¹⁴ M (0.000 000 000 000 01 M)

13 — 10⁻¹³ M (0.000 000 000 0001 M)

12 — 10⁻¹² M (0.000 000 000 001 M)

11 — 10⁻¹¹ M (0.000 000 000 01 M)

Milk of magnesia ⟶

10 — 10⁻¹⁰ M (0.000 000 0001 M)

9 — 10⁻⁹ M (0.000 000 001 M)

8 — 10⁻⁸ M (0.000 000 01 M)

Pure water ⟶ 7 — 10⁻⁷ M (0.000 000 1 M)
Cow's milk ⟶

6 — 10⁻⁶ M (0.000 001 M)

Cabbage ⟶ 5 — 10⁻⁵ M (0.000 01 M)

Tomatoes ⟶ 4 — 10⁻⁴ M (0.0001 M)

Soft drinks ⟶ 3 — 10⁻³ M (0.001 M)
Vinegar ⟶

2 — 10⁻² M (0.01 M)

1 — 10⁻¹ M (0.1 M)

Battery acid ⟶ 0 — 10⁰ M (1 M)

(M = Molar concentration in 1 L)

Fig. 3.12 The pH scale can be illustrated using some common household items. (Adapted from M. J. Pelczar, E. C. S. Chan and N. R. Krieg (1993) *Microbiology Concepts and Applications*. McGraw–Hill, New York.)

What is soil acidity?

We have already talked briefly about acidity. Instinctively we all know what is meant when we say something is acidic – we automatically think about lemon juice or vinegar. Lemon juice and vinegar are examples of organic acids. Their scientific names are citric acid and acetic acid. All acids have high concentrations of H^+ ions, and the greater the concentration of H^+ ions, the stronger the acid. We can describe the strength of an acid using the pH scale, which is a logarithmic scale expressing the concentration (more precisely, the activity) of H^+ ions in solution. We can illustrate the pH scale with some common household chemicals, as shown in Fig. 3.12.

Acidity determines a whole range of soil characteristics, such as the nature of the variable charge, nutrient availability, microbial activity and the release of certain toxins such as metals. Soil becomes acidic through several mechanisms, some of which are natural, like the acidity produced by the breakdown of soil organic matter, and some of which are caused by humans, such as pollution from industrial sources that then creates acid rain (see Chapter 7).

Measuring soil acidity not only tells us about the concentration of H^+ ions in solution but it can also indicate which cations are likely to be held in

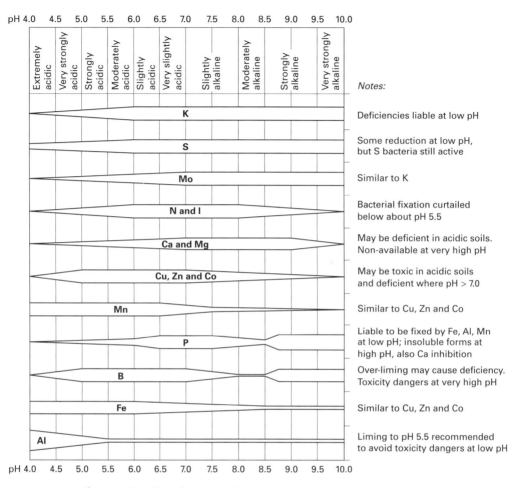

Fig. 3.13 The effect of soil pH on the availability of some common soil chemicals. (From J. R. Landon (1991) *Booker Tropical Soil Manual: A Handbook for Soil Survey and Agricultural Land Evaluation in the Tropics and Subtropics.* Longman Scientific and Technical, Harlow.)

the greatest concentrations on exchange sites. For example, acid soils often have high concentrations of exchangeable aluminium, unlike alkaline soils where exchange sites are more likely to be dominated by sodium (Na^+), calcium (Ca^{2+}) and magnesium (Mg^{2+}). Figure 3.13 shows how soil acidity can affect the availability of several plant nutrients.

Agricultural soils are particularly prone to gradual acidification because cations which counterbalance excessive acidity (sometimes called 'base cations'), such as Ca^{2+}, Mg^{2+} and K^+, are not returned to the soil naturally when the plant dies but are transported away after harvest. This results in the soil becoming increasingly acidic. In order to remedy this problem farmers need to understand how to manage soil acidity. The agricultural implications of acidity, as well as its management, are covered in Chapter 6.

Problems associated with acid soils

Acid soils can also be regarded as aluminium soils. At low soil pH (approximately 4), aluminium in the form of Al^{3+} comes into solution. This can cause plants nutritional problems because it can stick firmly to colloidal exchange sites, thus reducing the ability of the soil to retain base cations. High concentrations of Al^{3+} can further increase soil acidity because Al^{3+} ions attract hydroxyl ions, removing them from soil solution and thereby increasing the concentration of H^+ ions. A simplified version of Al^{3+} reaction with water is:

$$Al^{3+} + 3H_2O \leftrightarrow Al(OH_3) + 3H^+ \tag{6}$$

Aluminium is toxic to plants and animals because it can inhibit cell division and the elongation of plant roots. One of the diagnostic features used to spot aluminium toxicities is the development of brown stubby roots that are confined to the upper soil horizons. Stunted root development can be a serious problem as it diminishes the plant's ability to absorb water during the summer months. This reduces its ability to survive drought.

Since soil fertility is so closely linked with acidity, it is good soil management to measure it before crops are sown. However, this is not as simple as it may at first seem, because we first have to make a choice between measuring the concentration of H^+ in soil solution or measuring the concentration of H^+ in soil solution plus the stored acidity that is held on exchange sites. A simple analogy would be deciding on how to measure someone's wealth: do you simply count the loose change in their pocket, or take into account the value of their house, car and any other fixed assets they may have?

Whether we measure the acidity of the soil solution or the total stored acidity will depend on the purpose of the study. We can divide soil acidity into the following groups:

♦ active acidity;
♦ exchangeable acidity;
♦ residual acidity.

Active acidity is only the small fraction of the soil's total stock of H^+ ions that is currently in solution. Exchangeable acidity takes into account the H^+ and Al^{3+} ions held within the diffuse layer and able to move easily into soil solution. Residual acidity refers to the stock of H^+, Al^{3+} ions and $Al(OH)_x$ groups that are locked tightly onto the inner surfaces of colloidal material. Although residual acidity is not immediately exchangeable (just like the fixed assets in our wealth analogy), it can be released slowly. It is estimated that in clay soils residual activity can be up to 100 000 times greater than the active acidity.

Although soil acidity forms three separate pools, all the pools are interlinked. For example, if there is a small drop in the concentration of H^+ of

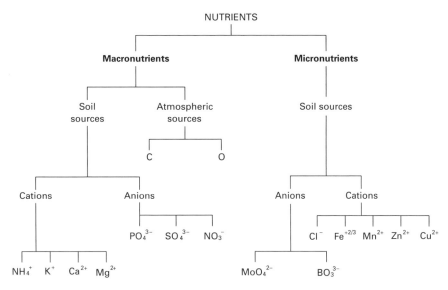

Fig. 3.14 Plants obtain their nutrition from a combination of soil and atmospheric sources. The nutrients can be divided into micro- and macronutrients.

the soil solution, other H^+ ions held on exchange sites will move into the soil solution, 'topping-up' the shortfall so that overall soil pH remains the same. The ability of the soil to adjust for changes in the ionic composition of soil solution is called 'buffering capacity'. It applies not only to acidity but also to nutrients held on exchange sites. All soils have some degree of buffering capacity; however, soils that have high CECs will have greater buffering capacities than soils with low CECs, such as sandy-textured soils. The practical implications of soil buffering capacity are covered in further detail in Chapter 6.

What are soil nutrients?

Nutrients can be defined as any substance used by an organism as food. We can divide soil nutrients into two groups, depending on their concentration in plant tissue. Macronutrients are found at concentrations in excess of 1000 mg/kg and micronutrients have concentrations below 1000 mg/kg. We will only consider the mechanisms that affect the availability of certain macronutrients in this section because all the other nutrients, including trace elements, are governed by similar processes. The division of nutrients is shown in Fig. 3.14 .

As carbon, hydrogen and oxygen are obtained predominantly from water and atmospheric sources they will not be considered further. We will consider the most important soil macronutrients, which are nitrogen (N), phosphorus (P) and potassium (K).

How is the availability of nitrogen, potassium and phosphorus affected by soil colloids?

Nitrogen

Despite being a very important plant nutrient, typically only 5% of soil nitrogen occurs in an inorganic form (not joined together with carbon and hydrogen atoms) in the soil solution, mainly as ammonium (NH_4^+), nitrate (NO_3^-) or, in very small quantities, nitrite (NO_2^-) ions. The majority of the nitrogen contained in the soil is held in the form of organic matter. This must be mineralized by soil micro-organisms before the nitrogen it contains is released for plant use. The biological mechanisms that govern this process will be given in more detail in Chapter 4.

Ammonium is rendered non-exchangeable or only slowly available by processes called 'fixation'. Because of its size, ammonium has the ability to penetrate the internal spaces that lie between individual 2 : 1 clay minerals in minerals such as vermiculite, illite and some forms of montmorillonite. Once held within the clay structure, ammonium only becomes available to plants very slowly. Up to 20% of the total nitrogen can be fixed in this way. Nitrate, because it is an anion, can be susceptible to leaching. In some agricultural areas where leaching losses are high, farmers have to employ careful management measures to control nitrate pollution (see Chapter 6).

Potassium

Potassium is an essential element in many enzymes; it also regulates the opening and closing of plant stomata, as well as being essential for photosynthesis and disease resistance. In soils, potassium is present as the cation K^+, which can be stored in three forms: K^+ in solution, exchangeable K^+, and non-exchangeable K^+. Potassium that is immediately available for plant uptake typically represents only 1–2% of the total soil stock. Exchangeable potassium (held on exchange sites) can readily exchange with the soil solution to buffer changes. The sum of the K^+ in soil solution and that held on exchange sites is the 'total available potassium'. Non-exchangeable potassium occurs fixed between individual clay minerals in 2 : 1 minerals such as illite. This reserve of potassium only becomes available for plants very slowly.

Plants use a lot of potassium, and may assimilate more potassium than they actually need. This tendency is called 'luxury consumption'. A combination of fixation and luxury consumption means that potassium can often be in short supply. The ability of a soil to buffer against potassium shortages will depend on its CEC and pH (see Fig. 3.13).

Table 3.2 The effect of pH on the form of soluble phosphorus.

pH	1	5	9	14
Form of P	H_3PO_4	$H_2PO_4^-$	HPO_4^{2-}	PO_4^{3-}

Phosphorus

Phosphorus is essential for photosynthesis, nitrogen fixation, crop growth, produce quality and root development. It is a component of both ATP (the molecule that drives metabolism) and DNA, the genetic template on which all life replicates.

Unlike potassium and ammonium, which are cations, in the soil phosphorus occurs as the phosphate anion. It can exist in several chemical forms, depending upon soil pH. Generally, plant-available phosphorus is highest in the pH range of 6–7. The commonest forms of phosphorus used by plants are $H_2PO_4^-$ and HPO_4^{2-}. The effect of pH on the phosphorus ion is shown in Table 3.2.

In soil the phosphorus anion rarely occurs in isolation with hydrogen ions. It is phosphorus reactions with other soil chemicals that reduce its availability to plants. The reactions that render phosphate unavailable occur when soil pH is either acidic or alkaline. For example in acid soils, iron and aluminium cations, coupled with hydrous oxides of aluminium and iron, can combine with the phosphate anion to produce a series of insoluble compounds. It is also believed that under acidic conditions phosphate anions can be fixed by certain clay minerals that further reduce its availability. In alkaline soils (>pH 7) a series of insoluble calcium phosphate compounds can form. This reduces phosphorus availability at the higher end of the pH scale.

In most soils the concentration of plant-available phosphorus is low, usually about 0.01% of the total amount of stored phosphorus. This is one reason why it is important to maintain the pH of agricultural soils at around 6–7, to ensure the maximum amount of plant-available phosphorus. Furthermore, the anionic character of phosphorus can be altered by other soil chemicals that lead to the formation of phosphorus compounds that are attracted directly to negatively charged colloidal surfaces (ligand formation). A combination of pH effects and ligand formation means that compared with other anions, such as nitrate, leaching losses of phosphorus tend to be small.

How can we describe a soil's ability to store and release nutrients?

As roots remove nutrients from the soil solution, losses are compensated for by buffering, as nutrients held on exchange sites are exchanged with

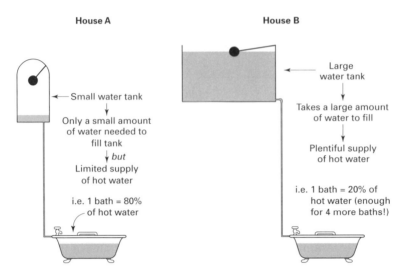

Fig. 3.15 Soil buffering capacity can be explained by using the analogy of two domestic hot-water tanks, one small, the other large, and the number of full bath-tubs each can fill.

the soil solution. In well-buffered soils, despite plant absorption and leaching losses, the concentration of nutrients in the soil solution is maintained. We can show buffering capacity using the analogy of two hot-water tanks. Look at Fig 3.15. It shows two houses: house A has a small hot-water tank while house B has a very large tank. Although the small hot-water tank does not take very much water to fill, it is soon emptied after one bath-tub. Although the large water tank in house B takes a long time to fill, once full it can fill several bath-tubs. In terms of buffering capacity, sandy soils have the equivalent to a small hot-water tank: easily filled but easily depleted. Soils with clay textures, on the other hand, have the equivalent to a large hot-water tank, in that plant absorption has only a small effect on the overall store of nutrients.

We can describe this process more scientifically using an intensity (I) versus quantity (Q) graph, sometimes referred to as a Q/I curve. The 'intensity' refers to the concentration of the nutrient in soil solution, whereas the 'quantity' is the concentration of the 'potentially labile' pool of nutrients held on exchange sites. In soils that are well buffered, nutrients can be released from exchange sites to compensate for declines in the nutrient concentration of the soil solution. However, in poorly buffered soils, even compensating for small declines in the nutrient concentration of the soil solution represents a large loss of stored nutrients held on exchange sites (just like the small hot-water tank). Figure 3.16 shows a couple of hypothetical Q/I curves, one for a sandy soil and the other for a clay-textured soil.

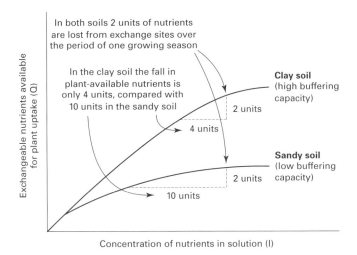

Fig. 3.16 The ability of a soil to supply nutrients can be explained using a Q/I curve. The steeper the curve, the better the ability of the soil to correct for changes in the nutrient concentration of the soil solution.

Essential points

◆ Soil acidity influences a whole range of biological and chemical reactions.

◆ Acidity is measured using the pH scale. Generally soil pH ranges from 3 to 9.

◆ Soil acidity can be classified as active, exchangeable or residual.

◆ We can divide soil nutrients into two groups, macronutrients and micronutrients, depending upon their concentration in plant tissue. The most important macronutrients are nitrogen, potassium and phosphorus.

◆ In most cases, the majority of soil nitrogen is contained within organic matter. The small fraction of inorganic nitrogen is mainly made up of ammonium and nitrate. Ammonium can become fixed by certain $2:1$ clay minerals whereas nitrate, because it is an anion, is susceptible to leaching.

◆ Potassium ions are attracted to negatively charged soil colloids. They can be held in soil solution, on exchange sites or fixed in a non-exchangeable form by certain $2:1$ clay minerals.

◆ Phosphorus occurs as the phosphate anion, but, unlike nitrate, potential leaching losses are reduced because it forms insoluble compounds when the pH is outside the narrow range of 6–7 and ligand bonds with other soil chemicals that modify its anionic character.

◆ The ability of the soil to supply nutrients can be described using a Q/I curve. The Q/I curve is a graphical description of the soil's buffering capacity with regard to changes in nutrient concentration.

Chapter Summary

Of the many fractions that constitute 'soil', in terms of soil chemistry clay minerals and humus are the most important. Their ability to affect the chemistry of the soil is related to their large surface area and electrostatic charge. Clay minerals are composed of sheets of silica tetrahedra and aluminium hydroxide octahedra. Some clay minerals, such as illite and kaolinite, have only external surfaces whereas others, such as montmorillonite, have both internal and external surfaces.

Soil organic matter consists of several chemical fractions. These can be divided into two groups: fresh residues and humus (which consists of products of advanced decomposition and strictly humic substances).

The electrostatic charge associated with colloidal material can be permanent (isomorphic substitution) or variable (through dissociation and protonation). The electrostatic charge is responsible for the attraction of ionic substances to the surface of the colloid. The surface chemistry of clay minerals can be divided into two sections. Layers closest to the clay surface are tightly held whereas other layers, at increasing distances from the surface, are held less tightly. The cation exchange capacity (CEC) is a measurement that can be used to describe the total amount of negatively charged exchange sites a soil has. Generally, soils with high concentrations of clay and organic matter have a high CEC.

Soil acidity can be categorized as being active, exchangeable or residual: of the three forms, residual acidity usually forms the greatest store but only slowly becomes available. Nutrients can be divided into two groups – macronutrients and micronutrients – depending upon their concentration in plant tissue. The most important macronutrients are nitrogen, phosphorus and potassium. The greatest store of nitrogen is found in the soil organic matter, where the nitrogen is in an organic form (bonded with carbon and hydrogen atoms). Organic nitrogen is mineralized to inorganic forms such as ammonium and nitrate. Ammonium can be fixed by certain clay minerals whereas nitrate is susceptible to leaching. Potassium cations can be fixed in a non-exchangeable form by some 2 : 1 clay minerals. Phosphate anions form insoluble compounds at low and high pH ranges. This reduces potential leaching losses. The maximum amount of plant-available phosphorus occurs in the pH range of 6–7. The ability of a soil to resist changes in the chemical composition of its soil solution is referred to as its buffering capacity. This tendency to buffer small changes in the availability of nutrients can be described using a Q/I curve.

4 Soil Microbes and Nutrient Cycling

Introduction

Imagine flying over a dense tropical rain forest. In your hand you hold a pair of binoculars; your task is to locate, count and then determine the ecological significance of ants on the forest floor. Impossible, you say – but the enormity of this task is of equal magnitude to the problems facing soil biologists. There are two reasons why the study of soil organisms is so difficult. First, soil organisms, especially micro-organisms, are a very diverse group – this makes identifying each and every individual a huge task. Secondly, many soil organisms live in close association with mineral and organic material, so that finding them, and then removing them for further study, can also be very difficult. Despite these challenges, soil biologists have shown great ingenuity over the years in unravelling some of the roles and interrelationships of this complicated subterranean world. This chapter will look at two main lines of enquiry:

---

1 **What types of organism are found in the soil?**
    How can we make sense of the biological complexity found in the soil?
    How can we group similar soil organisms together?
    How can we group soil organisms according to their ecology?

2 **How do soil microbes recycle nutrients?**
    What role do soil organisms play in nutrient cycling?
    Soil biota and nutrient cycling
    Carbon cycling and the role of inactive and active microbes
    The soil biota and nitrogen cycling
    Pool size, nutrient cycling and the concept of 'turnover'
    New horizons

---

1   What types of organism are found in the soil?

How can we make sense of the biological complexity found in the soil?

Global biodiversity estimates suggest that there are over 10 million different species, many of which occur in the soil. We could simply list all the different species, but this would do little to help us understand their biology and the ecological role they play. In order to understand this biological complexity, scientists have developed methods in which similar organisms are grouped together. This branch of science is called 'taxonomy'. By grouping similar organisms together we can start to understand the route by which they evolved and how they interact both with each other and with the wider environment.

The main problem with all taxonomic schemes is choosing which, out of the many characteristic features, should be used as the basis on which we decide that two organisms are similar. In older systems all organisms were simply divided into two broad categories, depending on whether they had broadly plant- or animal-like characteristics. This system was fine up to a point, when common sense rather than scientific enquiry was enough to say whether an organism had roughly plant or animal features, but not so good when biologists attempted to group micro-organisms together. This was because many microbes have a mixture of plant and animal characteristics.

In the 1970s, with developments in molecular biology, scientists started to develop molecular techniques to determine how organisms were related to each other. All organisms contain ribosomal ribonucleic acid (usually abbreviated to rRNA). If two organisms are compared, the greater the differences in rRNA the longer the period of time when both organisms shared a common ancestor. This has allowed scientists to look beyond simple physical differences to the genetic code of each organism, thus enabling organisms to be grouped together on a genetic, rather than simply a physical, basis. This has redrawn some of the old taxonomic divisions; the major one being the recognition that there are two major groups of bacteria, which are as distinct from each other as we are from bacteria. The division of life according to its genetic profile is shown in Fig. 4.1.

In addition to classification schemes based upon cellular characteristics, we can also group together organisms using other taxonomic schemes based upon how organisms obtain energy or carbon, or on their relationship to oxygen. All of these taxonomic schemes have the same objective, and that is to provide additional help in understanding the ecology of organisms. We will look at these 'ecological' approaches later, but we will begin by looking at how soil biologists have developed taxonomic systems

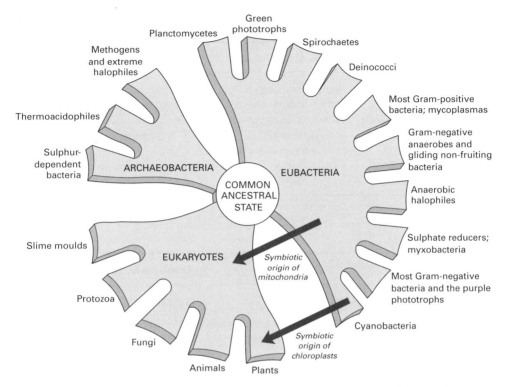

Fig. 4.1 Taxonomic systems group organisms together using a variety of characteristics. One of the modern systems is shown here, whereby organisms are divided into groups on the basis of their genetic profile. (Adapted from M. J. Pelczar, E. C. S. Chan and N. R. Krieg (1993) *Microbiology Concepts and Applications*. McGraw–Hill, New York.)

that use a combination of cellular characteristics and simple physical differences, such as size, to group similar soil organisms together.

How can we group similar soil organisms together?

Traditionally soil organisms have been divided into three size class ranges. The smallest of these is called the 'microbiota' and includes all organisms less than 200 µm. Bacteria, protozoa, fungi and algae are all members of the microbiota. Intermediate-sized organisms (200–10 000 µm) are called 'mesobiota'; this group includes organisms such as nematodes, rotifers, springtails and mites. Organisms over 1 cm in length are termed 'macrobiota'. This group includes earthworms, slugs, snails and many insect groups. We can use simple size differences to immediately divide soil organisms into three very broad groups; within each group organisms can then be classified on conventional taxonomic lines, as illustrated in Table 4.1.

We will now look briefly at which organisms are present in each size grouping, starting with the main soil organisms included in the microbiota.

Table 4.1 The use of size differences to group similar soil organisms.

Microbiota (<0.2 mm)	Mesobiota (0.2–10 mm)	Macrobiota (>10 mm)
Bacteria (Eubacteria and Archaeobacteria)	Nematodes (Eukaryotes)	Earthworms (Eukaryotes)
Cyanobacteria (Eubacteria)	Rotifers (Eukaryotes)	Large insects (Eukaryotes)
Slime moulds (Eukaryotes)	Springtails (Eukaryotes)	Snails (Eukaryotes)
Fungi (Eukaryotes)	Mites (Eukaryotes)	Centipedes (Eukaryotes)
Protozoa (Eukaryotes)	Small arthropods (Eukaryotes)	
Algae (Eukaryotes)		

Table 4.2 Approximate numbers of organisms (per gram) commonly found in the microbiota.

Organism	Estimated no./g
Bacteria (not including Actinomycetes)	3 000 000–500 000 000
Actinomycetes	1 000 000–20 000 000
Fungi	5 000–900 000
Algae	1 000–500 000
Protozoa	1 000–500 000

Source: M. J. Pelczar, E. C. S. Chan and N. R. Krieg (1993) *Microbiology Concepts and Applications*. McGraw–Hill, New York.

Microbiota (<0.2 mm)

Bacteria   One gram of soil may contain 200 million bacterial cells. In addition to their high individual numbers, the bacteria also represent a very diverse group of organisms, able to perform a vast range of different functions within the soil. Bacteria are unicellular, prokaryotic organisms that are usually between 0.5 and 1.0 µm in length. When exposed to substrate under the right temperature (10–25°C) and moisture conditions, bacterial cells can multiply rapidly. Table 4.2 shows the number of individuals per gram of soil for four groups of the microbiota.

   Bacterial numbers are particularly high in the thin film of substrate-rich soil that surrounds the plant roots ('rhizosphere'). Many soil bacteria are surrounded by a thick mucilaginous film that may protect the bacterial cell from hazards such as drying out and unfavourable shifts in pH. Bacteria tend not to occur loose in soil solution but are often anchored to the sides of soil particles. They attach themselves using a combination of gums and electrostatic charges. The reasons for this sedentary lifestyle may be related to the buffering capacity of clay minerals or the higher concentrations of nutrients found on clay surfaces. Whatever the reasons, the tendency for bacteria to be stuck firmly within the soil fabric is one of the

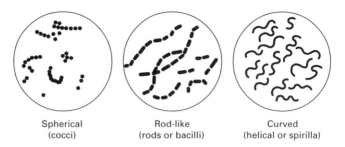

Spherical	Rod-like	Curved
(cocci)	(rods or bacilli)	(helical or spirilla)

Fig. 4.2 Some common bacterial shapes.

main reasons why they are so difficult to extract from the soil with their morphology and metabolism unchanged.

Microbiologists have traditionally recognized bacteria on several criteria, such as what reactions the organism can carry out and the shape of its cell. Bacterial cells are commonly rod-shaped (bacilli), round (cocci) or like a wavy line (spirilla). These groups are shown in Fig. 4.2. However, in soils this distinction can become blurred because some bacterial groups, such as *Arthrobacter*, can alternate between rods or spheres, depending on age and nutrition (rods are believed to indicate better nutrition). Other classification systems have used the ability of some bacterial cells to retain dyes. Cells that can be dyed are referred to as Gram-positive, whilst those that can not are referred to as Gram-negative (Gram is simply the name of the man who developed the stain). This distinction has been related to slight chemical differences within the cell wall of differing bacteria. Gram-positive bacteria have slightly thicker cell walls. Soil biologists will often refer to both the Gram stain characteristics and the cell shape when describing a bacterial cell. For example, *Pseudomonas* is a very common bacterial genus that can be described as having Gram-negative rod-shaped cells.

A group of bacteria called 'Actinomycetes' or filamentous bacteria are unusual in that, although classified as Gram-positive bacteria, they have many features similar to fungi, particularly the ability to produce thread-like extensions 10–15 µm long. Sometimes these extensions are permanent, as in *Streptomyces*, and other times only temporary, as cells alternate between rods, spheres and filamentous growth (e.g. *Actinomyces*). In some soils, particularly where the pH is above 7, Actinomycetes play an important role in the decomposition of resistant material. They are hardy organisms, able to withstand drought and defend themselves against competitors by synthesizing a wide range of antibiotics, some of which have been isolated and used in medicines.

Other bacterial groups, such as the cyanobacteria (previously called 'blue-green algae'), have some similarities with higher plants, the main one being the ability to use atmospheric carbon in a process called 'photo-synthesis'. However, unlike plants they are prokaryotic organisms, and so are classified as part of the Eubacteria. Some species also have the ability to

Table 4.3 Approximate biomass range of members of the microbiota in a grassland soil.

Organism	Biomass (t/ha)
Bacteria	1–2
Actinomycetes	0–2
Fungi	2–5
Protozoa	0–0.5

Source: K. Killham (1994) *Soil Ecology*. Cambridge University Press, Cambridge.

use nitrogen from atmospheric, rather than from soil sources. The ability to use atmospheric carbon and nitrogen has led to cyanobacteria playing an important role in the initial phase of soil formation: they are often referred to as 'free-living' or 'non-symbiotic' N fixers.

Fungi   If tissue mass (rather than individual cell numbers) is measured, fungi are the dominant microbial group in most soils, as shown in Table 4.3.

The fruiting bodies of some fungi are instantly recognizable as mushrooms or toadstools, but these structures represent only a small fraction of the total mass of the fungi, the main body of which lies concealed beneath the soil surface. Fungal tissues are made up eukaryotic cells, which form fine threads called 'hyphae' (2–10 µm diameter); these then coalesce to form larger filaments called 'mycelia' (Fig. 4.3). Fungal cells often contain more

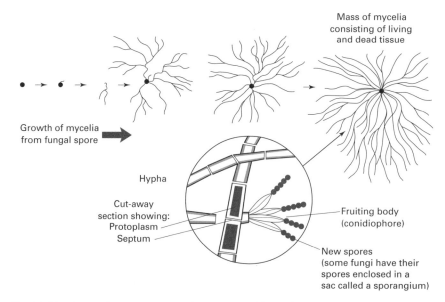

Mass of mycelia consisting of living and dead tissue

Growth of mycelia from fungal spore

Hypha

Cut-away section showing:
Protoplasm
Septum

Fruiting body (conidiophore)

New spores (some fungi have their spores enclosed in a sac called a sporangium)

Fig. 4.3 Fungi grow from spores to produce a mass of thread-like tissue called 'hyphae'. Not all the hyphae in the soil are biologically active. One of the problems in measuring the amount of live fungi in the soil is deciding which hyphae are living and which are dead.

Table 4.4 The effect of soil pH on the number (per gram) of bacteria and fungi.

Soil pH	Bacteria (millions/g)	Fungi (thousands/g)
7.5	95	180
7.2	58	190
6.9	57	235
4.7	41	966
3.7	3	280
3.4	1	200

Source: J. P. Martin and D. D. Focht (1977) Biological properties of soils, in Soils for Management of Organic Wastes and Waste Waters, ed. L. F. Elliott and F. J. Stevenson. American Society of Agronomy, Madison, WI.

than one nucleus; these nuclei can be moved to other cells along the hyphae network (rather like a tube train) via a central pore in the septum. Although 70% of the total weight of soil microbes is made up of fungal tissue, much of this mass is often dead hyphae lying in worked-out soil seams where the original source of nutrition has long since disappeared.

Fungi can survive harsh conditions by increasing the thickness of their cell walls to form 'chlamydospores', or by producing a structure called 'sclerotium', which consists of a mass of hyphae that can germinate to produce fruiting bodies. Other features that are characteristic of harsh conditions are 'rhizomorphs' and 'mycelial cords', both of which are made up of hyphae and are believed to help in the spread of nutrients and water.

Many fungi thrive under slightly acidic conditions, particularly where plant residues contain high concentrations of lignin. Some fungi groups, especially members of the genus Pisolithus, Poria and Amanita, play an important role in the decomposition of woody tissue, which is notoriously difficult to decompose because of its high lignin concentration. Table 4.4 shows the effects of soil pH on the numbers of bacteria and fungi.

Algae   Algae are single-celled organisms that are found as single cells, or grouped into colonies or as filamentous strands. They are photosynthetic organisms that are able to use atmospheric carbon like plants. Algae occur mainly in aquatic environments; they can sometimes cause water pollution when high growth rates can cause water to become deficient in oxygen. Some algae groups are widespread in soils, mainly occurring in the top horizons. Algae can play an important role as primary colonizers of rock surfaces in the initial stages of soil formation.

Protozoa   In the soil 'protozoa' is an umbrella term that includes three main groups of organism: flagellates (<5 μm diameter), amoebae (5–10 μm diameter) and ciliates (>20 μm diameter). These groups are illustrated

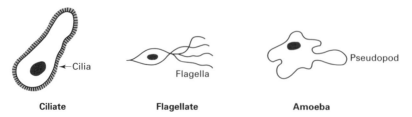

Fig. 4.4 Protozoa can be divided into three main groups, based on their mode of movement.

in Fig. 4.4. In one study, using an arable soil from the UK, a gram of soil was found to contain 70 500 flagellates, 1400 amoebae and 377 ciliates. Protozoan groups are distinguished on the basis of their locomotion. Flagellates, such as *Bodo*, move using whip-like extensions, while ciliates such as *Colpoda* move by moving bristle-like hairs. Amoebae such as *Naegleria* move by pushing out a long, finger-like projection of protoplasm, then moving the rest of the cell behind it.

Protozoa can be effective bacterial grazers. They are essentially aquatic organisms that are reliant on moisture films that surround soil particles for movement. Protozoa can survive adverse conditions, such as drought, by forming resistant structures called cysts. This allows them to suspend activity until conditions improve.

Mesobiota (0.2–10 mm)

Members of this group include small threadworms, called nematodes, a variety of small insects (arthropods), predatory worms (enchytraeids), rotifers, springtails and mites. In terms of nutrient cycling, nematodes are probably the most important members of the group. Commonly, they are between 0.5 and 1.5 mm long and, like the protozoa, are essentially aquatic organisms. Although some are serious soil-borne plant pests, the majority of soil nematodes feed upon members of the microbiota, helping to recycle nutrients.

Macrobiota (>10 mm)

This group includes all organisms larger than 1 cm in length. It includes enchytraeids, molluscs and larger arthropod groups. In terms of nutrient cycling and beneficial effects on soil structure, earthworms (e.g. *Lumbricus terrestris* or the European earthworm) are probably the most important members of the macrobiota. Earthworm numbers tend to be greatest in neutral soils (pH 7), that have high concentrations of organic matter such as grasslands. They are important because they play a crucial role in the physical breakdown of organic matter into smaller units (comminuting), thereby increasing its surface area and speeding up its decomposition by soil bacteria and fungi. Earthworms may also play a role as grazers, consuming other smaller organisms as they work their way through the soil.

How can we group soil organisms according to their ecology?

One of the problems with traditional taxonomic classification systems, which are based simply on the appearance of an organism or its genetic profile, is that they say little about its ecological role. For example, one group of bacteria, *Nitrobacter*, may look no different from countless other bacterial groups in the soil but it plays an important, highly specialized role by performing one of the steps that returns soil nitrogen to the atmosphere.

In addition to classifying organisms on their physical appearance and on how similar they are genetically, we can also group organisms together on the basis of how they obtain their carbon, energy, or on the basis of their relationship with oxygen.

Carbon source

All organisms require carbon. Carbon is used in two ways: firstly, it is used by many organisms as a source of energy and, secondly, it is used as the raw material from which cells are constructed. We can divide organisms into two groups, based on whether they obtain their carbon from the atmosphere (autotrophs), or through chemical compounds (heterotrophs).

Energy source

Other classification systems have considered how the organism obtains energy. Those that obtain energy through sunlight (like plants) are termed 'photosynthetic', while those that oxidize organic substrates are called 'chemotrophic', which includes all of the animal kingdom, fungi and many bacterial species.

We can combine these two systems to form a classification system which divides soil organisms into four groups: photoautotrophs, photoheterotrophs, chemoautotrophs, chemoheterotrophs, depending on how each organism obtains energy and its primary source of carbon. Figure 4.5 shows how we can use these divisions to classify soil organisms on an ecological basis.

Oxygen relationship

We can use respiration differences to divide soil organisms into three groups: those that require oxygen for metabolism (aerobes), those that require oxygen-free conditions (anaerobes) and, finally, organisms that are able to switch between both forms of respiration (facultative anaerobes).

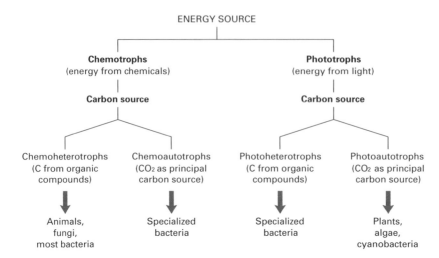

ENERGY SOURCE

Chemotrophs
(energy from chemicals)

Phototrophs
(energy from light)

Carbon source

Carbon source

Chemoheterotrophs
(C from organic
compounds)

Chemoautotrophs
(CO_2 as principal
carbon source)

Photoheterotrophs
(C from organic
compounds)

Photoautotrophs
(CO_2 as principal
carbon source)

Animals,
fungi,
most bacteria

Specialized
bacteria

Specialized
bacteria

Plants,
algae,
cyanobacteria

Fig. 4.5 Soil organisms can be grouped according to how they obtain carbon and energy. Grouping organisms on this basis illustrates their ecological role in the soil. (Adapted from B. N. Richards (1987) *Microbiology of Terrestrial Ecosystems*. Longman Scientific and Technical, Harlow. © Longman Group UK Ltd 1987, reprinted by permission of Pearson Education Ltd.)

Most soil organisms are aerobes, requiring oxygen to function. All of the mesobiota, macrobiota, the fungi and many bacterial groups are aerobes. Aerobic respiration is the most efficient respiration pathway, yielding far more energy per unit carbon metabolized than any other mechanism. This probably explains why, for most life forms, aerobic respiration is the norm. Facultative anaerobes include bacterial groups such as *Pseudomonas* and *Bacillus* that are able to switch to alternative forms of respiration, whereby other inorganic chemicals, such as NO_3^- or CH_4, are used instead of oxygen.

Finally, 'anaerobes' refers to those organisms that can function only in the *absence* of oxygen. These organisms thrive in boggy, water-saturated soils that are depleted of oxygen. They survive by reducing inorganic and organic substances; these may include sulphates, carbohydrates, organic acids and alcohols. When organic substances are reduced, the process is called 'fermentation'. Although the ability of these organisms to free themselves from the constraints of oxygen gives them a competitive advantage (particularly under waterlogged conditions) one of the disadvantages of anaerobic respiration and fermentation is that it yields far less energy per unit of substrate than aerobic respiration.

Taxonomic differences, size differences, the mechanisms used for obtaining carbon or energy and its relationship to oxygen all are ways of grouping similar organisms together. However, these schemes are not mutually exclusive, as they can be combined. For example, we may have a Gram-positive, rod-shaped bacterium that is chemotrophic, and able to function as a facultative anaerobe.

Essential points

♦ Soil organisms are very diverse, occupying a vast range of habitats.
♦ Traditionally they have been classified using size differences. The micro-biota (<0.2 mm) include bacteria, fungi, protozoa and algae; the mesobiota (0.2–10 mm) include nematodes, rotifers, insect larvae and springtails; and, finally, all organisms >10 mm are classified as macrobiota.
♦ Other classification systems have used the way the organism obtains energy or carbon, or its relationship to oxygen to group organisms.
♦ Classification systems are not mutually exclusive; it is quite easy to combine them to describe an organism.

2   How do soil microbes recycle nutrients?

What role do soil organisms play in nutrient cycling?

As plant and animal residues enter the soil they are colonized by a whole range of micro-organisms. As the residues are broken down nutrients are released and recycled through the soil. Initially, the easiest compounds to break down are attacked first, leaving the more resistant compounds to accumulate in the soil. Eventually, microbial waste products (microbial metabolites) and resistant plant residues combine to form soil humus in a process called humification (see Chapter 1). Although the combined weight of all the carbon contained in soil micro-organisms accounts for only 2–5% of the total mass of soil carbon, all the residues entering the soil must be broken up and recycled by micro-organisms before they can be used again by plants. Micro-organisms therefore perform a crucial function, recycling nutrients from plant residues back to the soil where they are available again for new plant growth.

Soil biota and nutrient cycling

Although taxonomic studies based on genetics, size and ecology can help separate soil organisms into convenient groupings, it is difficult to know what role each individual species plays in the long chain of interactions that makes up, let's say, a nutrient cycle or the break-up of a particular toxin. In order to understand how micro-organisms carry out major trans-formations we need to simplify the situation by studying them not as indi-viduals, or even as groups of similar organisms, but in bulk. This approach is analogous to looking at the ecology of the whole forest rather than the individual trees.

Instead of studying the thousands of different soil organisms separately, we can group them all together so that they form a single 'pool'. We can

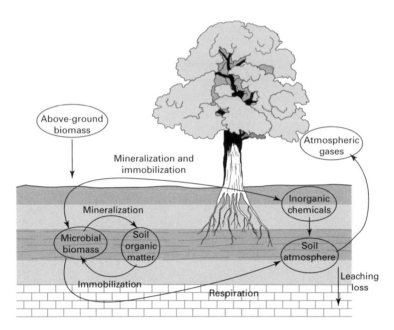

Fig. 4.6 A generalized nutrient cycle showing how we can divide the movement of nutrients into several interlinked pools.

also do this for soil organic matter, above-ground plants, and nutrients in solution, so that they form a series of interconnected nutrient pools that form the main components of a 'nutrient cycle', which traces the passage of a nutrient such as nitrogen from atmosphere to soil, and finally back to the atmosphere. The term 'microbial biomass' is used to describe the microbial pool. Figure 4.6 illustrates a generalized nutrient cycle that consists of several nutrient pools.

By simplifying the relationship between microbial biomass and soil processes, such as carbon and nitrogen cycling, we can start to investigate the contribution of micro-organisms to nutrient cycling. Here we are not interested in the millions of individual organisms carrying out countless tiny reactions, we simply want to find out how the size of the microbial biomass changes and how it relates to other factors, such as the amount of carbon or nitrogen in the soil.

We can adopt a similar approach to other pools such as soil organic matter, above-ground vegetation and carbon in solution. These pools are composed of many different fractions (for example, the soil organic matter pool consists of fresh residues and humified compounds). In order to study nutrient flows, a nutrient cycle can be divided into a number of pools. This helps simplify the situation, allowing scientists to study how the size of one pool affects the size of the others. Figure 4.7 shows the various pools of the carbon cycle. For clarity we have shown the cycle in several stages – in reality all the chemical transitions would be going on simultaneously.

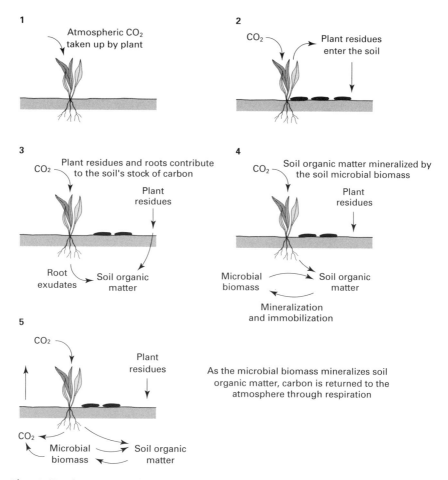

Fig. 4.7 For clarity we can show the carbon cycle in a series of stages. In reality all these processes would be happening simultaneously.

Recently, the need to understand the carbon cycle has taken on a new urgency with the rise in atmospheric CO_2, in a process known as the 'greenhouse effect'. Although the main factor that has caused this problem is the burning of fossil fuels, changing land use can also have a huge effect on the amount of carbon stored in the soil or released to the atmosphere. Some land uses, such as woodland, lead to an accumulation of carbon; others, like arable agriculture, can lead to carbon being lost from the soil. Changing land use can therefore exacerbate or help alleviate the greenhouse effect.

Long-term field trials represent some of the oldest experiments in the world. Although they were originally set up to measure the effects of fertilizers, their greatest contribution may still lie in the future, as scientists investigate ways in which rising concentrations of atmospheric CO_2 can be

 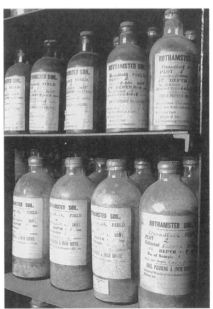

Fig. 4.8 Since the mid-19th century soil samples have been taken from long-term field trials at IACR-Rothamsted. Stored samples like these allow scientists to investigate how land-use change affects soil carbon stocks. These data are now being used to develop methods to offset the greenhouse effect. (Courtesy of the Photographic Department, IACR–Rothamsted Harpenden, UK.)

combated. The photographs in Fig. 4.8 show soil samples that have been stored for over 100 years at IACR-Rothamsted, site of the world's longest-running experiment. Stored samples such as these allow scientists to study how land-use change can affect soil carbon concentrations. The soil samples stored at Rothamsted are now of crucial importance, as scientists investigate how to combat global warming.

Carbon cycling and the role of inactive and active microbes

The amount of carbon contained within the microbial biomass is generally proportional to the size of the soil organic matter pool, which is in turn roughly proportional to the amount of carbon inputs. On any given soil type, generally the size of the biomass carbon pool declines in the order: grassland > forest > arable.

Even under productive grassland systems, when the size of the microbial carbon pool is measured, the estimated inputs of carbon are often insufficient to supply microbes with enough energy for both cell maintenance and reproduction. It has been known for some time that, unlike microbial cultures grown in the laboratory, soil organisms tend to have slow generation times. However, even when this is taken into account,

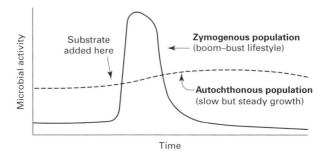

Fig. 4.9 Some soil organisms respond quickly to substrate; they have fast generation times and high death rates once the substrate is used. These organisms are termed 'zymogenous'. Other organisms have slow generation times and greater population stability; they are called 'autochthonous'. (Adapted from M. Wood (1989) *Soil Biology*. Blackie, Glasgow.)

the problem remains – that in most soils the estimated carbon inputs are insufficient to support the microbial biomass. One solution to this problem has been to assume not one, but a number of microbial pools within the microbial biomass, each with a different level of activity. In other words, it is assumed that some microbes are more active than others.

Research has suggested that only a fraction of the microbial biomass is active. The majority of soil microbes live in a reduced state of activity, requiring only small amounts of nutrients to maintain their cells. This concept of different microbial activities is not new; in the 1920s soil scientists suggested two types of organisms. One of these is fundamentally opportunistic, being able to respond quickly to fresh substrate. These organisms divide quickly and have rapid death rates once the substrate is utilized: they were called 'zymogenous'. The other group of soil microbes exhibited slow growth rates and slow death rates, and consequently, greater population stability: these organisms were labelled 'autochthonous'. It is assumed that under most circumstances the soil microbial biomass consists of a small but active zymogenous pool and a larger inactive autochthonous pool. Figure 4.9 is a schematic description of how each group responds to substrate. There is now experimental evidence to support this concept, though it is difficult to say what determines zymogenous or autochthonous characteristics. They may be related to taxonomic differences (e.g. fungi or bacteria) or simply to whether the microbe has access to substrate.

The soil biota and nitrogen cycling

Soils would soon become barren and infertile if it wasn't for the fact that nutrients are continuously recycled by the microbial biomass. In Chapter 3 we discussed inorganic nitrogen; now we must consider the most important reserve of nitrogen – the nitrogen contained in the soil's

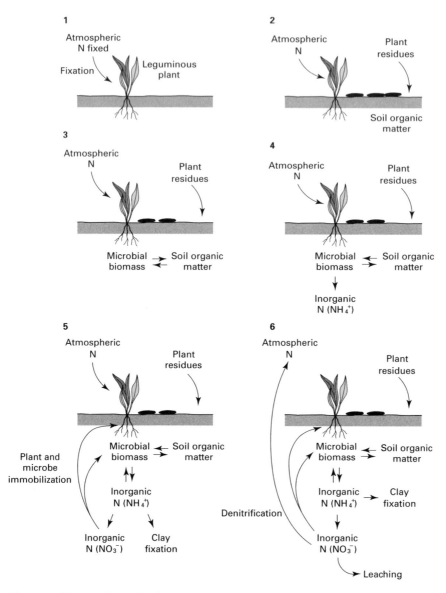

Fig. 4.10 Like the carbon cycle, the nitrogen cycle can be shown as a number of interlinked pools. Here we have built up the nitrogen cycle in stages for simplicity: in reality all the processes would be going on at the same time.

store of organic matter. The nitrogen cycle is controlled primarily by the interaction between the microbiota and the type of residues entering the soil. In a similar way to the carbon cycle, the nitrogen cycle can be shown as a series of interlinked pools (see Fig. 4.10). As with the carbon cycle, for simplicity we have shown the nitrogen cycle in a series of stages when in reality many of the reactions would be going on at the same time.

The pools in the N cycle

Atmospheric N   Approximately 80% of the air you are now breathing consists of N. Atmospheric N is present in the form of N_2; in other words, two nitrogen atoms joined together to form a 'diatomic' molecule. Although nitrogen is essential for life, before it can be used by organisms the dinitrogen molecule must be split open. The N_2 molecule is stable, requiring huge amounts of energy to break it apart (when this process is carried out commercially it requires in excess of 400°C at 200–350 bars pressure with a catalyst!). Given the high energy requirements the incorporation of N into organic molecules must have represented one of the greatest problems to face the evolution of life in its present form.

A solution to the problem was found when some organisms developed the ability to use atmospheric nitrogen: these organisms are called nitrogen fixers. They produce an enzyme – nitrogenase – which overcomes the problem of splitting the N_2 molecule apart. However, it is important that nitrogenase be kept away from oxygen and that there is a ready supply of carbon, because the process (even when an enzyme is used) is energetically very expensive for the N-fixing organism.

One of the most common bacterial groups of N_2 fixers, *Rhizobium*, overcomes these problems by entering into a symbiotic relationship (see Chapter 1) with certain plants known as legumes. *Rhizobium* cells occupy plant roots, forming characteristic nodules. In return for carbon, which the plant fixes using photosynthesis, rhizobial cells fix atmospheric N_2, which the plant then incorporates into its tissue. The ability of leguminous crops to fix atmospheric nitrogen (rather than extract it from the soil) is the reason why these plants are grown as green manures in areas of the world where the soils are nitrogen deficient. Nitrogen fixers provide the main route in nature by which atmospheric N_2 becomes incorporated into the various soil N pools.

Soil organic matter N   Most organisms cannot use atmospheric N_2, but have to scavenge for their nitrogen in the highly competitive soil environment. Once it is an inorganic form, nitrogen is used in a whole range of compounds, such as enzymes and amino acids. When the plant or microbe dies, its remains are rapidly mineralized by other microbes via a series of enzyme-mediated reactions. These 'catabolic' enzymes can occur within the cell (endocellular) or be excreted outside it (exocellular).

The contribution of residues to the soil's nitrogen supply depends on the quantity of residues, environmental factors (such as temperature and moisture) and, importantly, on the quality of the residue. Residue quality simply refers to how much and how quickly nitrogen will be released by mineralization. This is dependent on a combination of how easy the N-containing compounds are to mineralize and the ratio of carbon to

Table 4.5 Dry matter content of selected soil organic matter.

Material	C	N	C : N ratio
Bacterial cells	50	15	3.3
Fungi	44	3.4	12.9
Farmyard manure	37	2.8	13.2
Maize shoots	44	1.4	31.4

Source: D. S. Jenkinson and J. N. Ladd (1981) Microbial biomass in soil: measurement and turnover, in *Soil Biochemistry*, vol. 5, ed. E. A. Paul and J. N. Ladd. Marcel Dekker, New York.

nitrogen in the material. The last factor is usually referred to as the C to N ratio. It is calculated by dividing the amount of carbon (g) by the amount of nitrogen (g). We can use the C to N ratio of organic substrates to determine how much nitrogen will be mineralized. Obviously the measurement has widespread agricultural applications. Table 4.5 shows the C to N ratio of some common soil inputs.

Microbial N   In addition to mineralizing plant residues, soil organisms can also create short-term N shortages. For example, bacteria have an average C to N ratio of 5 : 1, whereas fresh straw residues can have values as high as 100 : 1. Micro-organisms must balance the concentration of nitrogen to carbon in their cells, so if a bacterial colony consumes a lot of carbon and expands, it must also find enough extra nitrogen to keep the C to N ratio of new cells the same as that of the original community. Study Fig. 4.11: what do you think will happen if fresh straw residues with a C to N ratio of 100 : 1 are ploughed into a soil where they are then mineralized by a bacterial community with a C to N ratio of 5 : 1?

If the bacteria consumed all the carbon contained in the straw (despite the fact that some of the carbon is lost as respiration) there remains a shortfall of nitrogen: this is usually met by using nitrogen from other sources, normally inorganic soil nitrogen. Effectively, this is what happens when fresh plant residues with high C to N ratios, such as straw residues, are ploughed back into the soil. As the residues are mineralized nitrogen is taken from the soil solution in a process called 'immobilization'. This can cause nitrogen shortages for plants.

It is for this reason that fresh plant residues, with their high C to N ratios, can cause temporary shortages of nitrogen. The reason why animal manure is often a better fertilizer than fresh plant residues is that it tends to have a lower C to N ratio. This is because some of the carbon has been lost through respiration as plant residues were digested and passed through the animal.

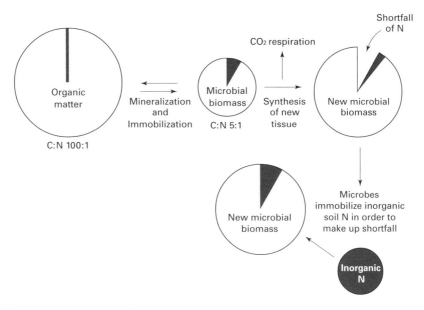

Fig. 4.11 The importance of the C : N ratio in the mineralization and immobilization of nitrogen by the soil microbial biomass.

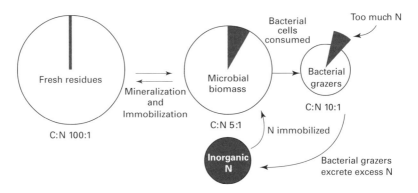

Fig. 4.12 The role of bacterial grazers in the release of nitrogen, immobilized by the microbial biomass, back to the soil solution.

Let us consider another example, shown in Fig. 4.12. A protozoan (C to N ratio = 10 : 1) squeezes itself into a water-filled pore to eat several bacterial cells (C to N ratio = 5 : 1): what do you think will happen to the nitrogen in this situation? As the bacterial cell is richer in nitrogen than the protozoan, the protozoan needs to rid itself of excess nitrogen. Animals usually do this by excreting ammonium (NH_4^+). The release of nitrogen by grazing faunal groups, such as protozoa and nematodes, may be an important mechanism in some soils for releasing immobilized nitrogen from bacterial and fungal cells.

Inorganic N   As plant residues and microbes are mineralized, nitrogen in the form of ammonium (NH_4^+) is excreted. When nitrogen is in this form, plants and microbes easily assimilate it. However, it is not only photo-autotrophs and chemoheterotrophs that use ammonium. A few bacterial species of chemoautotrophic bacteria oxidize ammonium to nitrate (NO_3^-) as part of their metabolism; this process is called 'nitrification'. Nitrification is a two-stage process; first ammonium is oxidized to nitrite (NO_2^-) by a number of specialist bacteria genera (*Nitrosomonas*); then nitrite is oxidized to nitrate by one specialist bacterial group (*Nitrobacter*). Although energy yields are low, the process is very rapid, so that in most agricultural soils most inorganic nitrogen is present as nitrate.

$$NH_4^+ \longrightarrow NO_2^- \longrightarrow NO_3^-$$
(Ammonium)   (Nitrite)   (Nitrate)   (7)

Nitrate can be used as a source of N by plants and microbes in a similar way to ammonium.

Back to atmospheric nitrogen   We must now finally consider how nitrogen is returned to the atmosphere. We have already said that in most neutral to alkaline soils much of the nitrogen initially mineralized as ammonium is quickly nitrified to nitrate. However, in situations where the soil microbial biomass remains active when oxygen concentrations are low, facultative anaerobes switch to anaerobic respiration, where inorganic chemicals such as nitrate and sulphate are used instead of oxygen. The process can be summarized as the following series of reactions:

$$\boxed{\begin{array}{c} NO_3^- \longrightarrow NO_2^- \\ \text{Solution} \end{array}} \longrightarrow \boxed{\begin{array}{c} NO \longrightarrow N_2O \longrightarrow N_2 \\ \text{Gas} \end{array}}$$
(8)

Nitrate is progressively reduced in a series of microbially mediated processes until we return to diatomic N_2 gas, which then is released back to the atmosphere, finally completing the nitrogen cycle. This process is called 'denitrification' and has major implications for nitrogen fertilizer losses (discussed further in Chapter 6).

Pool size, nutrient cycling and the concept of 'turnover'

In most soils, nitrogen is undergoing mineralization and immobilization at the same time but in different locations. Sometimes, where the ecosystem is undergoing change, there may be changes in the relative size of each pool. However, where the ecosystem has reached steady-state conditions (nutrient inputs balance outputs), despite the fact that the composition of nutrient pools is continually changing, there may be no overall change in the size of each pool. The constantly changing composition of pools is

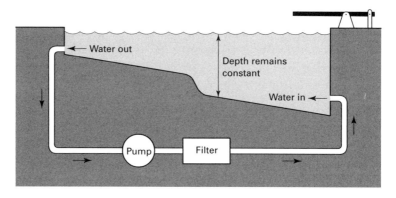

An ecosystem under **steady-state** conditions

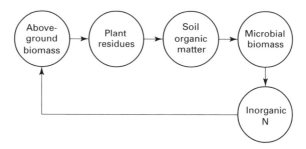

Although the composition of each pool is constantly changing, its size remains constant

Fig. 4.13 Mineralization and immobilization turnover (MIT) under steady-state conditions can be compared to a swimming pool's filtering system, whereby water is continuously being recycled but there is no change in the amount of water in the pool.

referred to as 'turnover' or mineralization–immobilization turnover (MIT). The concept of turnover can be illustrated if you imagine a swimming pool (shown in Fig. 4.13) where the water is continuously being changed as water is sucked out and filtered before being returned to the pool. Despite the fact that the water is continuously being changed, there is no overall change in the amount of water in the pool. Similarly, although the size of the microbial biomass may not alter over the year, new biomass is being continuously synthesized as other microbes die and are mineralized.

New horizons

Recently the emphasis in soil biological research has switched to assessments of soil quality (or health) using bio-indicators. Some scientists believe that certain microbial factors, such as specific bacterial or fungal species or even the microbial genetic diversity of the soil, may provide a simple method of measuring the quality of the soil. Measuring the genetic diversity of the soil may also be useful for other purposes. Traditionally, attention has always focused on the genetic diversity of large ecosystems such as tropical rain forests: less is said about the vast pool of genetic diversity that lies beneath our feet every time we stroll across a lawn or ploughed field. So far we have barely scratched the surface of what potential reactions soil organisms may carry out.

Essential points

♦ In order to simplify the ecological complexity of the soil we can group all the microbes, organic matter (residues and SOM), gases and inorganic compounds together in a series of pools.
♦ We can use the concept of interlinked pools to explain the movement of nutrients through the soil in a nutrient cycle.
♦ In the case of the microbial biomass, only a fraction of the microbes are active – these are called zymogenous – whereas inactive microbes are called autochthonous.
♦ Nutrient cycling is governed by the interaction between the microbial biomass, the quantity of mineralizable organic matter and the quantity of the organic matter, especially its C : N ratio.
♦ The C : N ratio of residues is a useful measurement to determine if nitrogen will be immobilized or mineralized.
♦ Nitrogen, once mineralized to ammonium, can be nitrified (converted to NO_3^-). This process is called nitrification and is performed by chemo-autotrophic bacteria in a two-stage process.
♦ In conditions where there is a combination of good carbon supply and low oxygen levels nitrate can be converted to N_2. This process is called denitrification and is the mechanism by which nitrogen is returned to the atmosphere.
♦ The exchange of nutrients within a pool is referred to as turnover. Under steady-state conditions the influx of new material and the removal of old material is not reflected in changes in pool size. Where the ecosystem is under change, there may be changes in the relative size of each pool as the soil approaches a new equilibrium.

Chapter Summary

Soil organisms can be classified in a number of ways. Traditionally they have been grouped into three size classes and then into further classes using general taxonomic differences. However, soil organisms can also be classified on an ecological basis, such as how they obtain energy or carbon, or their relationship with oxygen. These taxonomic systems do not need to be mutually exclusive; they are simply tools to help understand a complex system by grouping similar organisms together. In order to study nutrient cycling, scientists have simplified the process by combining all the various microbial, organic matter and inorganic fractions together in a series of interlinked pools. This approach enables scientists to concentrate on the ways in which the size of one pool alters in relation to others under a range of environmental conditions. Although the composition of each pool is continually changing, as new material enters and old material leaves, when the ecosystem is at equilibrium, turnover may not always be reflected in changes in pool size.

5   Soil Survey, Classification and Evaluation

Introduction

In order to understand how soils change, scientists have developed survey methods and statistical techniques to analyse soil properties. Changes in soil properties may be associated with different management practices (e.g. the conversion of forest to arable land), with time (e.g. where an arable soil is left to convert slowly to forest) or with location (e.g. the same soil type at different positions along a hill slope). Although chemical and biological measurements are usually made in the laboratory, the quality of the final data is highly dependent on how the soil was sampled in the field. If the samples were collected carelessly, using poor survey methods, the data will be useless regardless of the precision of later laboratory measurements. In order to produce high-quality data, scientists need to know how to sample in the field.

In addition, scientists often need to know how different soils will behave in relation to management changes such as drainage, tillage, reforestation and cultivation. It would be impossible to test every single soil individually, so, to save time and money, surveyors have attempted to group similar soils together. Similar soils can be grouped together according to their physical, chemical or visual characteristics. This allows a generalized set of characteristics to be assigned to each soil type. However, unlike plants and animals, soils do not fall into discrete units but change gradually: this makes it difficult to say where one soil type ends and another one starts. Despite these difficulties, soil scientists have succeeded in defining distinctive soil groups that form the basis of soil classification.

We will begin by looking at how surveys are conducted, particularly at how the samples should be taken. Then we will look in more detail at surveys that analyse changes, followed by descriptive 'general' soil surveys that aim to group similar soils together. The final section of the chapter will look at soil maps and how they are used in agriculture. This chapter will take four lines of enquiry:

1   How do we carry out a soil survey?

What are we trying to do when we conduct a soil survey?

It is sometimes useful to clarify the meaning of an everyday word because we all tend to use words such as 'survey' without actually thinking about their meaning. According to my dictionary, survey means to view, scrutinize, inspect, examine, measure and map. We could quite easily have used these definitions as subheadings for this chapter. Soils can be viewed, scrutinized and inspected at a number of levels. Here we have divided soil surveys into two types. First, there are those studies that measure changes in one or several selected soil properties. These might be pH, nutrient concentration, water content or any other soil characteristic. The aim of these surveys will normally be to measure changes over time, location or the effects of various treatments that have been applied to the soil; sometimes the survey will involve testing a hypothesis. Secondly, there are surveys that seek to obtain a general impression of the soils of the area; these types of survey will normally be carried out by a soil surveyor who will describe the soil in general terms using a classification system. Field observations are then transformed to produce a soil map that can be used by a wide range of users for a variety of purposes.

In both cases, the scientist or soil surveyor will need to be clear about how to sample the soil. Successful sampling aims to take the samples in such a way that, despite the fact that only a fraction of the soil in the area is sampled, the measurements and descriptions from these samples can be used to explain the soils over the whole of the sample area. In order to do this it is important to plan carefully how the samples are taken, where they are taken, and how many are taken.

Taking the samples

Initially you might think we could just wander around the site collecting samples at will, but in reality you often find it is quite difficult to be completely random in your selection of sampling sites (it is all too easy to miss sampling those far-flung, inaccessible areas). In order to ensure complete coverage of the site a more systematic approach to sample collection is needed; in other words we need to design a sampling strategy. We will look at two ways of taking samples: grid and free survey sampling techniques.

Grid survey

A grid survey is conducted using intersecting transects which are plotted across the whole survey area: soil samples are then taken at the intersections. As shown in Fig. 5.1, the pH of each sample can be measured and the values plotted on a map. Similar values can be joined together with lines called 'isolines' to create boundaries showing varying high and low pH. Grid surveys are particularly useful where there are no clues to changes in soil properties, such as clear changes in vegetation or relief.

Free survey

Free surveys use features such as changes in vegetation and topography to target the sampling strategy. For example, if our farm was not flat and featureless but divided into flat productive arable land (with regular liming), forested areas and waterlogged hollows, we could guess that the main changes in soil properties would occur between these three different land uses. If we were measuring pH, we might expect the variation in pH within each land use to be small in comparison with differences between land use. Free surveys enable the surveyor to target the sampling strategy. As there is no need to plot grids, the time and the number of samples can sometimes be reduced (see Fig. 5.2).

Once the surveyor has identified the likeliest soil boundaries, samples can then be collected at regular intervals along transects plotted across

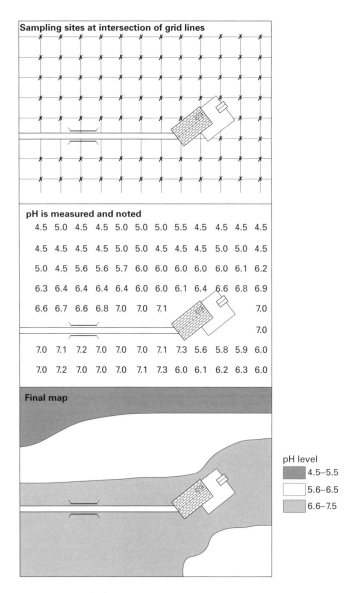

Fig. 5.1 Mapping the pH of a farm using grid survey methods.

each of the subsites. If time and money allow, many transects can be plotted. In Fig. 5.3 each transect has been labelled: A, B and C for the forest site; D and E for the arable site; and F and G for the waterlogged hollow. Sometimes these transects can be 'W' shaped so that the whole area can be covered in one transect. Unlike a grid survey, which shows both the pH of each area as well as the gradual change across the site, a free survey such as this will simply show the mean pH for each area and its variation along each transect.

Fig. 5.2 Plotting a transect. The farm where this survey was conducted was divided up into several sample areas. For each area transects were then staked out before samples were taken at 10 m intervals.

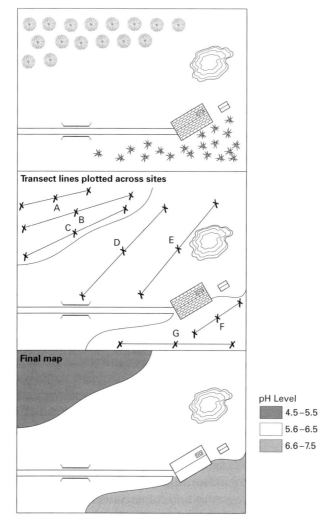

Fig. 5.3 Mapping the pH of a farm using free survey methods.

Essential points

♦ Soil surveys can be used to verify, inspect, measure and finally map the distribution of soil.

♦ Surveys can be conducted at a number of levels. These can range from the small-scale local surveys, which may be interested in only one or two soil properties, to general soil surveys that aim to show the world distribution of soil.

♦ It is important to sample the soil correctly so that the small amount of soil tested accurately reflects the soil in the field.

♦ Samples can be collected using grid or free survey techniques.

♦ Sometimes the aim of the survey will be to investigate changes in selected soil properties; at other times it will be to describe the soil in general terms.

2   How do we conduct a soil survey that measures change?

Main objectives

These surveys have clearly defined objectives because there is often a well-defined purpose for the results. They are mainly conducted by scientists interested in looking at how soil properties vary in relation to other factors, such as changes in land use or fertilizer applications. The main decisions the scientist has to make often relate to the choice of sampling strategy. As with all surveys, it is important to be clear, not only about how to collect the samples, but also about how many samples need to be taken. For example, a detailed study over a large area will require a lot of samples, taken at close intervals, whereas a survey that simply aims to gain a general impression may only require a few widely spaced samples.

Statistical analysis of the data

In some cases it may be required to test whether there has been a significant change in the soil properties under investigation. In order to illustrate this, let us return to our hypothetical farm pH survey we discussed in the last section, but this time let us say that the farmer has added lime to half of the forest site to raise its pH. He now wants to know if the lime applications have been effective. The scientist's task is to answer this question. This type of survey is not a simple mapping exercise; it requires statistical analysis so that the conclusions can be reached with a degree of certainty.

What the scientist needs to ensure is that differences in pH are really due to lime amendments. The sampling techniques are the same as used

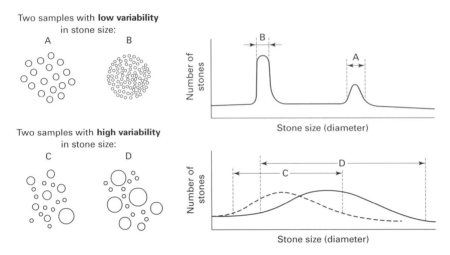

Fig. 5.4 We can use the standard deviation to describe sample variability. Here we have illustrated the concept using four piles of stones. Piles A and B have low sample variability, whereas piles C and D have great variability.

in mapping exercises, but in order to test any variations in pH statistically we need to measure how the pH varies *within* each area before comparing variation *between* areas. It is the ability to compare variation within the limed and non-limed areas that enables us to say with confidence whether the lime applications had a significant effect on soil pH.

Figure 5.4 illustrates this concept, using four piles of stones. In two of the piles, A and B, the size of stones is largely uniform; in the other piles, C and D, stone size is highly variable. If we plot a graph of stone diameter against number of stones, we produce two differently shaped curves. Note how the shape of the curves from piles A and B are steep whereas those from C and D are long and flattened. We can express variation in stone size by calculating the 'standard deviation' of each pile of stones. The standard deviation is a basic statistical measurement that indicates how widely individual values are dispersed from the mean. In a statistical sense we could say that piles A and B have little variation, which could be expressed by a small standard deviation, whereas piles C and D have high variability and a large standard deviation. Standard deviation can be calculated using a basic scientific calculator (look for the symbol $x\sigma^{n-1}$).

Although the data can be shown graphically using a map, the usual method is to express the differences between each site using a bar chart, so that each bar corresponds to the mean pH of each site. We can show how variable the measurements are (as in the stone example) by expressing the variation as small lines on the top of each bar. These are called 'error bars', and are calculated using the standard deviation. These allow the reader to immediately assess the spread of the data and estimate whether

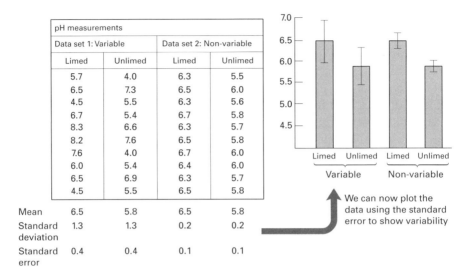

pH measurements			
Data set 1: Variable		Data set 2: Non-variable	
Limed	Unlimed	Limed	Unlimed
5.7	4.0	6.3	5.5
6.5	7.3	6.5	6.0
4.5	5.5	6.3	5.6
6.7	5.4	6.7	5.8
8.3	6.6	6.3	5.7
8.2	7.6	6.5	5.8
7.6	4.0	6.7	6.0
6.0	5.4	6.4	6.0
6.5	6.9	6.3	5.7
4.5	5.5	6.5	5.8

Mean	6.5	5.8	6.5	5.8
Standard deviation	1.3	1.3	0.2	0.2
Standard error	0.4	0.4	0.1	0.1

We can now plot the data using the standard error to show variability

Fig. 5.5 Two sets of data are shown. In set 1 the data are highly variable, in set 2 the data are non-variable. Although the mean is the same, the spread (or variability) of the data is different for each set. This is illustrated by the lines at the top of each bar.

any differences are likely to be statistically significant. They can be calculated using the following formula:

$$\text{Standard error} = \frac{\text{standard deviation}}{\sqrt{\text{number of samples}}} \tag{9}$$

Consider the data in Fig. 5.5. It shows two sets of pH measurements from our farm, one from the non-amended area and the other from the limed area. In the first example the data are very variable while in the second data set variation is small. Now compare how the amount of variation makes no difference to the overall mean but does affect the standard deviation and the standard error. See how the standard error when expressed as a line on top of each bar clearly shows the spread of values within each data set.

The standard deviation and standard error are two of the simplest methods whereby we can describe data numerically. In order to test whether differences are actually significant, statistical tests such as analysis of variance (sometimes referred to as ANOVA) and the Student's *t*-test can be used. These tests allow the scientist to say with varying degrees of confidence whether the differences in soil properties under investigation are due to treatment differences, location, time or any other variable that has been identified as the source of change. A fuller description of statistical techniques, and tests detailing how and under which circumstances they should be applied, can be found in any simple statistics book.

Essential points

♦ Soils can change in relation to management practices, time and location.
♦ Data can be expressed in the form of maps, tables and graphs. When data are represented (plotted) in a graph, it is important to show error bars, which indicate the variability of the data.
♦ Statistical tests such as analysis of variance and Student's *t*-test can be used to test whether changes are significant.

3   How do we conduct a soil survey that describes the soil in general terms?

The problem of soil classification

In some respects measuring changes in soil properties is straightforward because the scientist has a clear idea about the objectives of the survey and the information that needs to be supplied. In the last example it was soil pH but it could quite easily have been soil textural analysis, organic matter, nutrient supply or any other soil property. Descriptive soil surveys can be more difficult because the main objective is to describe the soil in general terms so that a range of users are satisfied. These might be agronomists interested in land quality, geographers seeking to understand how a soil has formed, or atmospheric scientists investigating soil gas emissions. The aim of soil classification is to combine similar soils into classes that are meaningful to a diverse range of end-users. We have already covered this topic briefly in Chapter 1, when we outlined some of the general properties associated with four soil profiles that we simplistically labelled podzols, gleysols, ferralsol and histosols. We now need to look in a bit more detail at how soils are classified and mapped for general-purpose soil maps.

Before a classification scheme can be developed we need to be clear about three questions:
♦ What soil characteristics are we most interested in?
♦ Can these characteristics be used as the basis of a classification system?
♦ How can we divide soils into discrete units when they naturally form a continuum?

The first question stems from the varied purposes for which the soil map can be used. As we said earlier, an agronomist may have different interests to a geographer. The second question highlights the fact that some soil properties may be useful, but only to a limited range of end-users. If these measurements require expensive, time-consuming laboratory analysis, they do not represent the most practical choice on which to base the classification system (this point is applicable to many of the engineering and biological properties of the soil). Finally, the third question, 'how do we group soils into discrete units?' is important because it is central to the problems of soil classification, and this is that soils do not occur as discrete units, but as types that change gradually across the landscape. Whilst biological taxonomists, for the most part, deal with 'black and white' issues centred on plant and animal species, soil scientists have to contend with numerous shades of grey.

Despite these problems, soil scientists have developed classification schemes for soils. In early classification systems soils were grouped together if they formed under similar conditions, whereas modern systems have tended to place more emphasis on classifying soils on the basis of observable and measurable properties.

Early systems of soil classification – how they worked and why they were replaced

Archaeological evidence taken from early agricultural settlements suggests that even primitive farmers could distinguish good soils from bad ones. One of the first attempts to classify soil types using soil properties, rather than how suitable a certain soil was for crop growth, was developed in Russia towards the end of the 19th century. Scientists such as Dokuchaiev and Sibirtzev believed that soil formation could be linked with environmental factors, such as climate and parent material. On this basis soils were grouped into three broad classes: Zonal, Intrazonal and Azonal soils that were characteristic of certain soil-forming factors:

♦ Zonal soil: these form in environments where climate has played the dominant role in soil formation.

♦ Intrazonal soil: these form where soil formation is dominated by factors other than climate. One example is where soils have formed directly on limestone. In cases such as these, the main soil characteristics are often more influenced by the chemical nature of the parent material (the limestone) rather than by climate.

♦ Azonal soil: these form where soil development is incomplete because the parent material was only deposited recently. One example is where soils have formed on recent alluvium.

The main problem with the early Russian approach is deciding which environmental factors have played the dominant role in soil formation. This is particularly true where the soil properties have been influenced by past environments. For instance, in the Russian system we need to establish the importance of climate, but soil profiles may develop over thousands of years, and over several major climatic changes. Consequently, there can be problems in ranking environmental factors in order of importance. This means that in many cases soil development is open to alternative interpretations. More recent systems of classification have adopted an alternative approach in which observable, measurable properties are used as a basis for soil classification rather than interpretations of soil formation. These classification systems are sometimes referred to as 'morphological systems'.

What soil properties do we use to classify a soil?

Although morphological systems have the advantage that no reference is made to soil formation, there is still the problem of deciding which soil characteristics to use in order to formulate each soil class. All the major soil classification systems use a well-defined list of horizon characteristics in order to classify the soil. Before the surveyor begins the survey he or she needs to know which classification system is going to be used and what soil properties need to be measured or described. A generalized list of some soil properties that are used in several classification schemes is shown in Table 5.1.

Table 5.1 A check-list for soil survey and related data collection.

1. Site details	
Location and topography	Grid reference and aerial photos
Land-forms/hydrology and geology	Relief and elevation; local drainage, main rock formations/parent material, depth below soil
Vegetation and land use	Description of dominant species, size and density of natural vegetation, cropping history, soil/ land relationships
2. General soil information	Surface stoniness, rock outcrops, erosion hazard, presence of salt, details of survey methods
3. Soil morphology	Horizon depths and thickness, moisture status, colour and mottling, structure, stone content, root size and distribution, horizon boundaries
4. Field tests	Bulk density, pH, carbonate content, textural analysis (finger method)
5. Laboratory tests	Textural analysis (mechanical analysis), carbon and nitrogen concentration, water content at field capacity
6. Additional and specialized information	Climatic data, annual precipitation, mean monthly temperatures, clay mineralogy, moisture release data, soil mechanical properties, toxic materials, microbial biomass, microbial diversity, organic matter fractionation

Source: J. R. Landon (1991) *Booker Tropical Soil Manual: A Handbook for Soil Survey and Agricultural Land Evaluation in the Tropics and Subtropics*. Longman Scientific and Technical, Harlow. © Longman Group UK Ltd. 1991, reprinted by permission of Pearson Education Ltd.

The US system of soil classification

The system developed by the United States Department of Agriculture (USDA) is called 'Soil Taxonomy'. It is one of the most comprehensive systems of soil classification and consists of a number of levels of classification. Classification systems that use this level-based approach are called 'hierarchical'. Soil Taxonomy draws distinctions between various soil types using the physical and chemical properties of the horizons that make up the soil profile. These are termed 'diagnostic horizons'. Each diagnostic horizon has a list of characteristics it must possess in order to be included within a group. When the characteristics of each horizon have been noted the surveyor can then begin the job of classifying the soil.

Soils can be described using six levels of classification, the broadest being 'order', followed by 'suborder', 'great group', 'subgroup', 'family' and 'series'. All we will consider in this chapter is the broadest class – the order. This is because at the lower levels of the system descriptions become very detailed, as soil classes are divided and then subdivided. For example, the USA alone has something in the order of 17 000 different soil series. The main soil orders are shown in Table 5.2.

Table 5.2 Soil Taxonomy (1975): soil orders.

Soil group	Description
Entisols	Soils with weak horizon development; examples include disturbed soils or soils formed on alluvium
Vertisols	Soils containing more than 30% clay which show cracking when dry
Inceptisols	Soils with little profile development; examples include young soils
Aridsols	Soils with high salt concentrations. These often contain salt crystals when dry. Examples include soils from arid regions
Mollisols	Soils containing epipedons with both high organic matter concentrations (>1%) and high base saturations (50%)
Spodosols	Soils containing high concentrations of organic matter, iron and aluminium. In some cases a fragipan (iron pan) may be present. In older systems this order is usually referred to as a podzol
Alfisols	Soils which show evidence of lessivage and have high base saturation levels (>35%)
Ultisols	Soils which show evidence of lessivage and have low base saturation levels (<35%)
Oxisols	Soils which have been subjected to long periods of weathering and are rich in aluminium and iron oxides. Typically they have a low CEC and only traces of weatherable material. This order is most commonly typified by the soils of the humid tropics
Histosols	Soils rich in organic matter

Soil Taxonomy has been criticized as being too inflexible, because it does not allow for the natural variation found as soils change gradually across the landscape. The system is also quite rigid in that the absence of one diagnostic feature is enough to exclude a soil from a particular group. This has led to the concern that surveyors may be tempted to ignore problematic field observations in order to be able to categorize all soils neatly into the groups defined by Soil Taxonomy. The other criticism is the need for laboratory measurements in addition to field observations before a soil can be classified. This increases the time and expense of the survey, and may be impractical in some areas of the world that do not have laboratory facilities.

What are the alternatives to Soil Taxonomy?

Although the US system has been used internationally, many countries retain their own national systems. This is because some soil types are only found in a few areas of the world, so they are unimportant globally, despite the fact that they may be very important at the national level. For example, compared with Soil Taxonomy, the Soil Survey for England and Wales gives greater attention to gleyed features (if you are unsure about gleyed soils see Chapter 1), emphasizing the crucial role of drainage in the successful cultivation of UK soils. In England and Wales drainage is regarded as being so important agriculturally that it forms the basis of one of the 'major groups'. In Soil Taxonomy, however, drainage characteristics are only considered at the suborder level of classification. The major groups of the Soil Survey for England and Wales are shown in Table 5.3.

Although many national systems are ideally suited for describing the soils found in their national boundaries, the proliferation of classification schemes sometimes makes it difficult to relate the soil types of one classification with those of another. The UK represents an extreme example of the proliferation of local schemes in that all the regions (England and Wales, Scotland and Northern Ireland) have their own classification systems.

Moves to unify soil classification systems

In 1961 a joint scientific advisory panel made up of representatives from the FAO (Food and Agriculture Organization), Unesco (United Nations Educational, Scientific and Cultural Organization) and ISSS (International Society of Soil Science) met to discuss preparing a 1 : 5 000 000 soil map of the world. The main aims of the project were to catalogue the world's soil resources and to put in place a generally accepted framework of soil classification. The scheme was not intended to replace national classification schemes, but to formulate a set of broad soil classes that could be used to describe the main soils of the world. In 1966 general agreement was reached on the formulation of the soil classes. This enabled scientists to start

Table 5.3 Soil Survey for England and Wales: major groups.

Soil group	Description
Gley soils	Soils which are subjected to long periods of waterlogging. Horizons can be grey-coloured with mottling
Lithomorphic soils	Soils with shallow profiles, characterized by organic horizons overlying bedrock
Pelosols	Soils with high clay concentrations which crack when dry
Brown soils	Well-drained soils with no gleyed features above 40 cm. Good agricultural soils which are divided further into subgroups on the basis of degree of lessivage and mottling, below 40 cm
Podzolic soils	Soils generally formed under acidic environments showing accumulations of organic matter and accumulations of iron and aluminium oxides. There may be the development of an iron pan in the B horizon
Man-made soils	Self-explanatory, examples include soils formed from remediation processes following mining and quarrying
Peat soils	Soils rich in organic matter, with organic matter accumulations at least 40 cm thick

correlating data from existing soil surveys, topographical data, vegetation data, aerial photographs and ground surveys to map the world's soils.

The project was completed in 1974 and bore some similarities to Soil Taxonomy. In the original FAO system the world's soils were classified into 26 major soil groupings that were subdivided into a total of 106 soil units. As in Soil Taxonomy, soil classes were defined on the basis of a set of diagnostic horizons, each having a set of observable, measurable properties. The project proved so successful that in 1988 the classification scheme was updated. The original two levels of classification were split into three levels, which were termed 'major soil groups', 'soil units' and 'soil subunits'. The need for this greater detail was prompted by the increasing use of the classification system in developing areas, especially in Africa. The latest amendments to the FAO system were made in 1998 with the creation of the World Reference Base (WRB) for soil resources. The WRB provides a way of correlating the many national classification schemes. There are two levels of classification, consisting of 30 reference soil groups and 170 possible subunits. Table 5.4 shows each reference group with a brief description of the profile's main properties. It should be noted that the description given here presents a highly simplified version of the 1998 FAO system. For a more detailed explanation see the chapter by Otto C. Spaargaren (2000), 'Other Systems of Soil Classification', in the *Handbook of Soil Science*, ed. M. Sumner.

Table 5.4 Reference Soil Groups used by the FAO (1998) *World Soil Resources Report 84.*

Soil group	Description
Histosols	Soils that have a peat layer or 'H' horizon >40 cm deep
Cryosols	Soils that have permanently frozen horizons (below 0°C for two or more years) within 100 cm of the soil surface
Anthrosols	Soils that have horizons strongly influenced by human activity such as cultivation (Hortic horizon), irrigation (Irragic horizon) or manure incorporation (Terric horizon)
Leptosols	Soils with hard rock within 25 cm of the surface or which overlie materials with >40% calcium carbonate within 25 cm of surface or contain <10% fine earth to a depth of 75 cm or more (weakly developed soils)
Vertisols	Soil that has a vertic B horizon (a clay-rich, self-turning horizon, >30% clay) >50 cm deep and within 100 cm of the soil surface
Fluvisols	Soil having formed on recent alluvial deposits, having a fluvic horizon (a dark-coloured horizon, usually resulting from pyroclastic deposits) within 25 cm of the surface, continuing to depths greater than 50 cm
Solonchaks	Soils that have surface or shallow subsurface horizons >15 cm deep that are enriched with soluble salts. Common in soils forming from recent alluvial deposits (soils of salty areas)
Gleysols	Soils which have evidence of gleying within 50 cm of the surface
Andosols	Soils that have vitric horizons (>10% volcanic glass or other volcanic material) >30 cm deep or andic horizons (weathered pyroclastic deposits) within 25 cm of the soil surface (volcanic soils)
Podzols	Soils that have a spodic B horizon (subsurface horizons containing illuvial organic matter and or aluminium and iron). Bleached surface horizons and an iron pan at depth are also usually present
Plinthosols	Soils with a plinthic horizon (iron-rich horizon >15 cm with >25% plinthite) within 50 cm of the surface which harden when exposed
Ferralsols	Ferric B horizon (distinctive red mottling >15 cm deep), highly weathered with high concentrations of iron and aluminium
Solonetz	Soils with a natric B horizon (a dark-coloured horizon >7.5 cm deep with clay enrichment and a high concentration of exchangeable sodium)
Planosols	Soils that have E horizon resulting from prolonged exposure to stagnant water within 100 cm of surface, often marked with an abrupt change in textural properties
Chernozems	Soils that have a mollic A horizon (dark-coloured, well-structured surface horizon with a high base saturation) to a depth of at least 20 cm
Kastanozems	Soils with a mollic A horizon >20 cm depth coupled with concentrations of calcium compounds within 100 cm of the soil surface
Phaeozems	All other soils which have a mollic A horizon

continued

Table 5.4 (*cont'd*)

Soil group	Description
Gypsisols	Soils that have gypsic (concentrations of calcium sulphate) horizons within 100 cm of the surface or concentrations >15% gypsum over 100 cm
Durisols	Soils having a duric horizon (cemented silica) within 100 cm of the surface
Calcisols	Soils that have calcic horizons (discontinuous concentrations >15% of calcium carbonate) within 125 cm of the soil surface
Albeluvisols	Soils that have an argic horizon (B horizon showing signs of clay enrichment) with an irregular upper boundary (sometimes referred to as 'tonguing')
Alisols	Soils that have an argic B horizon with a CEC >24 $cmol_c$/kg clay and a base saturation <50% within 100 cm of soil surface (soils with high concentrations of aluminium)
Nitisols	Soils that have a nitric B horizon (a clay-rich horizon with >30% clay mainly consisting of 1 : 1 minerals) >30 cm deep with a CEC <36 $cmol_c$/kg with no evidence of clay lessivage detected using thin sections within 100 cm of the surface
Acrisols	Soils that have an argic B horizon with a low CEC (<24 $cmol_c$/kg). These soils are characterized by acidity
Luvisols	Soils with an argic B horizon with a CEC >24 $cmol_c$/kg with illuvial accumulations of clay
Lixisols	All other soils having argic B horizons within 100–200 cm of the surface
Umbrisols	Soils having umbric horizons (thick, dark-coloured base, poor surface horizons)
Cambisols	Soils with a cambic (evidence of change or alteration) or mollic horizon over a subsoil with a base saturation <50% in the top 100 cm or an andic, vertic or vitric horizon starting between 25 cm and 100 cm depth
Arenosols	Weakly developed coarse-textured soils
Regosols	All other soils

Essential points

♦ General soil surveys attempt to categorize soil into types that have a similar collection of physical, chemical and biological properties.

♦ Soils are classified on the basis of the diagnostic horizons that occur in the profile. Each soil class has a well-defined set of characteristic horizons (called diagnostic horizons).

♦ Although there are many international schemes, such as the USA's Soil Taxonomy, many countries have retained their own systems of soil classification.

♦ The FAO World Reference Base for Soil Resources provides a way of enabling comparisons to be made between the soil classes of national classifications.

4   How do we produce a map from soil descriptions made in the field?

The formation of mapping units

The final output of most general soil surveys is a soil map depicting where each soil type occurs. Soil maps are produced at various scales. In the field, profiles are usually described and classified at the very detailed series level, but it may be impractical to depict every soil series on a small-scale map (imagine illustrating every soil series found in the USA with a different colour). Sometimes, in order to produce a soil map that is informative but doesn't suffer from information overload, we need to group soil types (as mapped in the field) into broader groups. These groupings are referred to as 'mapping units'.

One type of mapping unit is formed by grouping soils that commonly occur together into one unit. These units are called 'soil associations'. In other cases, areas of the soil map depicted as one soil type may in reality contain a mixture of soils. Mapping units that contain more than one soil type are called 'soil complexes'. The map legend will often list the average percentage composition of the main soil types that make up each association or complex. Once a soil has been described in detail in the field the surveyor is free to produce a range of soil maps suited to a variety of different purposes.

When organizing the final output, the surveyor's job is to present the information to best suit the interests of the end-user. Mapping units allow survey data to be expressed in such a way that the most relevant observations are clearly presented. Figure 5.6 shows two maps of the same area. In the first map the surveyor has simply listed the major soil types in the area. In the second, the surveyor has chosen to express the potential erosion risk of the area.

How can we use the soil map in land evaluation?

One special use of the soil map is land evaluation. This is an example of how information gathered in a general soil survey can be interpreted for a specialist purpose, which is evaluating land for either general agricultural production or for specific crops. Land can be evaluated in two ways:
♦ land capability systems;
♦ land suitability systems.

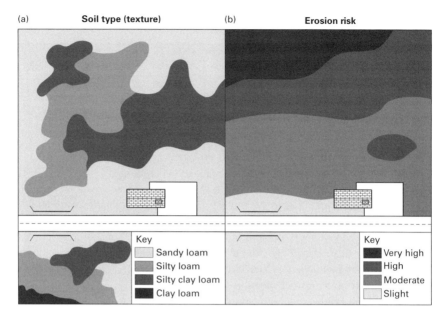

Fig. 5.6 The same location depicted using two maps. The first map (a) shows soil type, the second (b) erosion risk.

Capability systems grade land on its potential suitability for general agricultural production. Land that has the greatest agricultural flexibility is awarded the highest grading whereas land with the greatest limitations is given a lower grade. In suitability systems there is a change of emphasis, whereby land is evaluated with reference to a particular crop or agricultural practice. This approach narrows down the land-use considerations, allowing more specific decisions to be made about how a given area should be managed. We will begin by looking at land capability systems.

Land capability systems

The USDA developed one of the first land capability systems, whereby land was divided into two major groups. First, there is land that has permanent limitations and, secondly, there is land with only temporary agricultural limitations which, if managed correctly, has agricultural potential. Land with temporary limitations was further divided into several categories, ranging from land with few limitations to land with severe restrictions. Originally the US system was designed to grade land on the basis of its erosion risk; however, it has now become more of a general system of land evaluation. The limiting factors for each class are shown in Table 5.5.

Table 5.5 Classes in the USDA land capability system.

Class	Description
Class I	Land with few limitations. Land in this class can support a wide range of crops with maximum returns from investments such as fertilizers
Class II	Land that may require moderate conservation measures to combat erosion, for example soil on gentle slopes
Class III	Land with limitations, both in terms of the choice of crops that can be grown and the degree of soil management required
Class IV	Land with severe limitations, both in terms of choice of crop and further increases in the need for careful conservation measures and skilful soil management
Class V	Land that has limitations other than erosion risk. For example, it may be in an area of high rainfall where excess water is a problem, or there may be soil factors such as high stone content. The cultivation of these soils is not usually possible
Class VI	Land with similar limitations to V but less easily managed. Land in this class is usually used for pasture, woodland or wildlife
Class VII	Land with limitations that cannot be rectified by management. Potential use for land in this class is woodland, recreation or conservation
Class VIII	Land with no commercial agricultural potential; for example, old mine workings, rock outcrops, sandy beaches

The UK has a similar system whereby land is divided into seven classes (Table 5.6). In the UK system less attention is paid to erosion hazard and more to climatic limitations, particularly annual precipitation. Classes 1–4 are considered suitable for all kinds of agriculture; 5 and 6 suitable for a range of purposes with the exception of arable production; whereas land awarded a classification of 7 is not suitable for agriculture. Tables 5.6 and 5.7 give descriptions of each land class, accompanied by the shorthand notation that indicates the nature of the main limiting factor. Now study Figure 5.7. It shows an area with a wide range of land classes with the main limiting factor in each case shown in parentheses.

Land capability systems have several advantages, the most obvious being the distinction they draw between land which is suitable for agriculture and that which is not. The disadvantage with capability systems is that they tend to be over-simplistic, in that in some circumstances low-grade land may be ideal for certain specialist crops. In order to grade land on the basis of individual crop requirements we need to use a suitability system.

Table 5.6 UK land capability classes.

Class	Description
Class 1	Land with no limitations. Easily working soils that retain moisture, i.e. deep fine sandy loams
Class 2	Land with minor limitations, which result in reductions in the choice of crops that can be grown. The land may also require some careful management, i.e. good soils on gentle slopes or soils that require some drainage measures
Class 3	Land with moderate limitations requiring careful management, i.e. soils prone to drought or waterlogging
Class 4	Land with moderate to severe limitations that requires very careful management; for example, soils prone to serious flooding or drought that only produce crops with poor yields
Class 5	Land where the agricultural limitations restrict its use to pasture
Class 6	Land with limitations so severe that its use is confined to rough grazing
Class 7	Land with no agricultural potential

Table 5.7 Symbols denoting the nature of the limitations.

Symbol	Description
s	Soil main limiting factor
g	Gradient, where the degree of slope reduces the soil's agricultural potential
w	Wetness
c	Climatic limitations such as altitude
e	Denotes a susceptibility to erosion

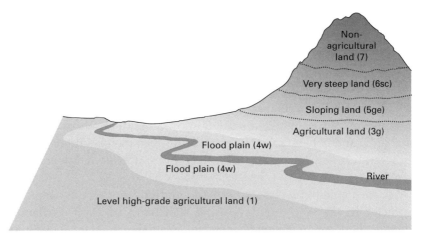

Fig. 5.7 Here a hypothetical river valley has been graded on its agricultural potential using the UK capability system.

Table 5.8 Nutritional requirements and salt and acid tolerance for a range of crops.

| Crop | Nutrition (kg/ha) | | | Tolerance | |
	N	P_2O_5	K_2O	Acidity	Salt
Barley	173	650	173	very sensitive	tolerant
Cotton	195	70	135	moderate	tolerant
Wheat	190	76	216	slight tolerance	slight tolerance
Cabbage	293	70	271	very sensitive	sensitive

Table 5.9 FAO soil suitability classes.

Order	Class	Description
Suitable	S1	Land which is highly suitable, with no major limitations
	S2	Land which is moderately suitable
	S3	Land which is only marginally suitable
Not suitable	N1	Land which is currently not suitable
	N2	Land which is permanently not suitable

Land suitability systems

Rather than dividing land on the basis of its general agricultural potential, suitability systems focus on the requirements of specific crops or agricultural practices. Farmers ultimately want to grow crops that will maximize their income but require minimum inputs. Using these criteria, land can be graded with reference to its suitability for certain crops. The FAO *Framework for Land Evaluation* (1976) is a suitability system that was originally created for land assessment in developing countries. Table 5.8 shows the nutritional requirements and acidity and salt tolerance of several crops.

We can use data like this to assess how suitable a particular soil is with reference to a specific crop or agricultural practice, such as irrigation. The soil of an area can then be graded into two main orders of suitability: suitable (S) and not suitable (N). These orders can then be divided further into several classes, depending on the degree of suitability. These are shown in Table 5.9.

With each suitability class we can add an indication of the main limiting factor to the land-use class using shorthand notation. Lower-case letters, such as (e) for erosion hazard, (m) for moisture, (r) for rooting conditions and (s) for salt problems, can be used to denote the nature of the limitation. These can be combined with upper-case letters to show the recommended management option: (D) for drainage, (I) for irrigation, (T) for terracing and so on. For example, let's say we wanted to grow rubber in an area that was relatively dry. Since rubber plants require a lot of water, the land is considered only marginally suitable for rubber production (S3m). However, let us now assume that there is access to a plentiful supply of clean irrigation water (I) that could be used to increase the land suitability classification to S1. We could write this information down as S3m/I/S1.

How are soil maps likely to be used in the future?

Advances in computer technology have revolutionized our ability to access, interpret and distribute soils data. The acknowledgement that the earth functions as a closed system, where changes in one area have a knock-on effect on other areas, has meant that soil maps are now being used outside the narrow confines of soil science and land evaluation.

One important driving force behind these developments is the need to understand how changes in land use can affect the mineralization of soil organic matter. As the soil organic matter is one of the major stores of carbon, a rise in the rate at which it is lost from the soil has serious implications for climate change and the greenhouse effect, by increasing the amount of solar heat stored within the atmosphere. A wide range of scientists, including meteorologists, agronomists, hydrologists, and policy makers working for international bodies such as the UN, have started to recognize the global ecological importance of soil. This recognition has led to the development of several global databases that include a soil component.

The Soils and Terrain Digital Data Base Project (SOTER) is one of the most important collections of electronic soil data and builds upon the experience gained through the development of the FAO *Soil Map of the World*. In 1995 the map was digitized, enabling it to be manipulated using computer software. This expanded its applications, so that what started out as a straightforward inventory of world soils became an electronic application that delivered a database that could be combined with other data sets, to provide a tool for linking soil characteristics with the wider environment.

Although SOTER is a new electronic system for storing information, it uses many of the concepts that we have already discussed. The mapping units used in SOTER can be divided into two sections. The first has been labelled 'Terrain Data' and stores spatial information relating to location (i.e. where in the world the soil is) using GIS (Geographical Information System). The other section is called 'Attribute Data' and includes information on the geology, the landforms (geomorphology) and the soils of the area. Figure 5.8 shows how SOTER mapping units are constructed.

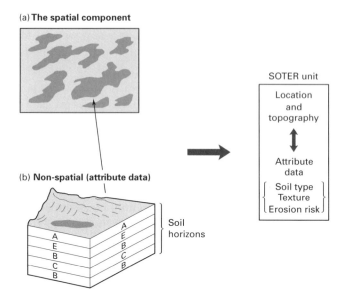

Fig. 5.8 The construction of SOTER mapping units.

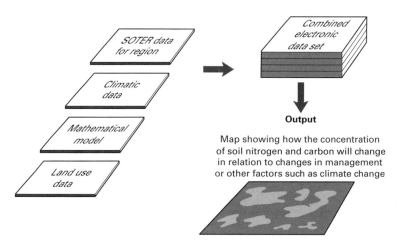

Fig. 5.9 SOTER mapping units can be combined with other data sets and mathematical models. Techniques such as these are being used to investigate the effects of land-use change on soil properties. These studies may have important implications for the management of global warming.

By combining SOTER data with other electronic data sets we can investigate a whole range of land-management options such as erosion control, food production and land classification. Each data set can be combined to form a series of interlinked 'layers' that allow soil characteristics to be seen in the context of the wider environment (Fig. 5.9). One of the latest developments has been to combine soil data with mathematical simulation models that investigate the effects of a range of land management options on soil carbon and nitrogen.

Essential points

♦ Mapping units can be used to help us simplify soil descriptions made in the field. They can be used to express a range of soil properties.

♦ Soil maps can be used to evaluate land for agricultural purposes.

♦ Land capability systems grade land in relation to its general agricultural potential.

♦ Suitability systems grade land with reference to a specific crop or agricultural purpose.

♦ The ability to digitize soil maps now means that soil data can be combined with other sources of information, such as land-use data, relief maps and mathematical simulation models. This represents a major step forward in the use of soil maps.

♦ The Soils and Terrain Digital Data Base Project (SOTER) combines location, soil, geological and landform data in one electronic mapping unit. It is currently being used to investigate how land-use change can affect the emissions of greenhouse gases from soils.

Chapter Summary

The evaluation of soil properties can be carried out on several scales, ranging from local surveys at the farm level to international global mapping exercises. In simple surveys only one or two soil properties may be measured. This may be simply to map their distribution or to assess whether there are significant differences between locations. These surveys are fairly simple to conduct because the surveyor's observations are limited to one or two well-defined properties. In general soil surveys the whole soil profile is assessed. Soils can be divided into broad groups using the physical, chemical and biological properties of the horizons that make up the soil profile. A number of classification schemes have been developed throughout the world. Two commonly used international systems are Soil Taxonomy, developed by the USDA, and the FAO system, which was developed originally as a legend for the 1974 *Soil Map of the World*. The World Reference Base (WRB) for soil resources aims to provide a common source of reference, so that soils classified using national systems can be related to each other. The end product of most general soil surveys is a soil map. The distribution of soils is usually simplified using mapping units. Examples of mapping units include 'soil complexes' and 'soil associations'. The soil map can be used in land evaluation. Capability systems evaluate the land in relation to its general agricultural potential, whereas suitability systems grade land with reference to a specific crop or agricultural purpose. Recent advances in computer technology have meant that soil maps can be combined with other electronic data sets, such as climate change and weather data, to aid our ability to understand how soils interact with the wider environment.

6　Soils and Agriculture

Introduction

Faced with an extremely waterlogged site full of deep puddles, bogged-down machinery and poor crop growth, what would you do? Despite the fact that we have now covered several of the main areas of soil science, we still have not looked at the practicalities of soil management. This chapter will focus on what to do when there is too much water, not enough water, compacted horizons, high acidity and several other common soil management issues.

Soil science has its origins in agriculture. Before the development of agricultural practices, nomadic tribes learnt how to manage their environment using fire, both to clear woody vegetation and as a way of hunting animals. Later, probably in response to population pressures, the growth of some food plants was promoted and other species were suppressed. The development of the first agricultural-based societies, around 10 000 years ago in the Middle East, led to a fundamental shift in the way people viewed their environment because permanent settlement meant private property and the notion of land ownership, class differences, urbanization and, finally, industrialization.

Although early farmers were expert at grading soils according to their agricultural properties, they did not understand the reasons why some soils tended to be more fertile than others. It was not until work was published by Theodore de Saussure (1804) and Justus von Liebig (1840) on plant physiology, coupled with the development of the first scientific agricultural trials by J. B. Boussingault (1834) in Alsace and John Bennet Lawes (1843) in England, that the role of soil factors in plant growth was fully appreciated.

The main task of the farmer is to create the ideal conditions for plant growth by managing soil properties. The root zone forms the main interface between plant and soil. Plant roots supply air, water and nutrients as well as providing anchorage. Soil conditions that interfere with these functions will reduce plant growth. This chapter will take three lines of enquiry:

1 How can we optimize the physical condition of the soil for plant growth?
   Problems with too much water: drainage
   What types of drainage are available?
   Problems with not enough water: storing what we have and applying
     extra
   Maximizing the storage of available water: water conservation
   Adding extra water: irrigation
   Potential hazards associated with irrigation
   Changing the physical state of the soil: cultivation
   The importance of soil moisture in cultivation

2 How can we optimize the chemical condition of the soil for plant growth?
   Providing the plant with the right chemical environment
   How can we control acidity by applying lime?
   Liming materials and application rates
   Supplying nutrients: fertilizers
   Types of fertilizer and the timing of applications
   Fertilizers and the environment: how can we minimize the pollution
     risk?

3 How can we optimize the biological condition of the soil for plant growth?
   Using natural biological cycles to maintain soil fertility
   Organic manure: benefits and potential problems

1   How can we optimize the physical condition of the soil for plant growth?

Problems with too much water: drainage

Soil moisture gains and losses can be compared to a bank account. During the wet winter months the soil's water 'account' is in credit as it soaks up water. In summer the soil loses water, often becoming overdrawn as water losses exceed inputs. In the winter, with the onset of cool weather, the moisture debt is repaid. Problems occur when the amount of water remains high during the growing season so that plant growth is disrupted. Figure 6.1 shows some of the problems associated with poor drainage.

Waterlogging occurs when the inputs of water, through rainfall, ground water or overland flow exceed outputs, through evaporation and drainage. If the soil remains wet for long periods plant roots will suffer oxygen shortages; this will be coupled with the accumulation of metabolic waste products, some of which are toxic. Under very cold conditions waterlogging is not necessarily a problem because root activity tends to

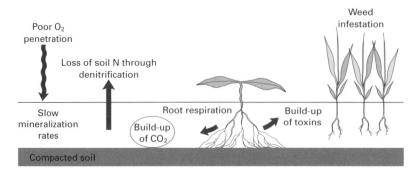

Fig. 6.1 The effects of poor drainage on plant growth and soil properties.

be very low. Problems occur when soils become waterlogged when the temperatures are high enough to allow biological activity. In the UK it is estimated that half of all farmland is so wet that it requires some form of drainage.

The effects of poor drainage not only include the obvious ones such as areas of free-standing water, but also more subtle ones such as plant leaves with dull mottled colours, weed problems, and patchy and N-deficient crops. It is not only places such as the UK, with its wet climate, that require drainage; in dry areas where crops are irrigated, good drainage schemes need to be installed to ensure irrigation water can drain away.

Drainage is a financial investment. The farmer will want to be confident that the substantial financial outlay of a new drainage scheme will be recouped with better crop growth. A cost–benefit analysis of any drainage scheme will have to take account of the following factors:
♦ the value of the crop;
♦ crop tolerance to waterlogged conditions;
♦ the harvest requirements of the crop (e.g. some crops require heavy harvesting machinery that needs to be supported by firm, well-drained soil).

What types of drainage are available?

Let us assume that it is economically worthwhile to install drainage. There are a number of options and the decision on which to use will be based upon the severity of the drainage problem and the budget available for rectifying it. When waterlogged conditions are due to the slow movement of water through the soil, the drainage scheme should aim to ensure that the soil reaches field capacity within 48 hours. Where the drainage problem is due to a high water table, drainage should aim to lower the water table so that it lies 1–2 m below the soil surface (Fig. 6.2).

We can divide drainage schemes into surface and subsurface measures.

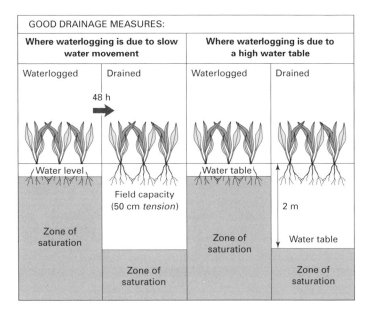

Fig. 6.2 The diagram shows the aim of drainage in soils where waterlogging is due to either slow water movement or a high water table.

Surface works (ditches)

Have you ever noticed the way some fields are surrounded by deep ditches? In many cases these ditches are installed for drainage. They are relatively easy to construct and can be used in isolation or as part of other drainage works. Although they offer easy entry for excess water, they effectively take land out of production. Ditches are usually dug to a depth of 1.0–1.5 m and are effective for draining areas that have high water tables, particularly where the rate of water movement through the soil is reasonably fast (medium- to coarse-textured soil). The sides of the ditch can be particularly prone to slippage, so may need stabilizing with regular maintenance to ensure the ditches remain at the correct depth and free from weeds and other obstructions (Fig. 6.3).

Subsurface works

In soils where the movement of water is slow we have to consider sub-surface drains to supplement ditches. To be effective, drains must be placed below the water table or zone of saturation. There are a number of types of subsurface drain. Three of the most popular ones are tile drains, mole drains and subsoiling.

Tile drains consist of ceramic pipes laid end to end. Water percolates through the saturated soil and into the pipe, which discharges it into drainage ditches at the field border. Perforated plastic piping can also be

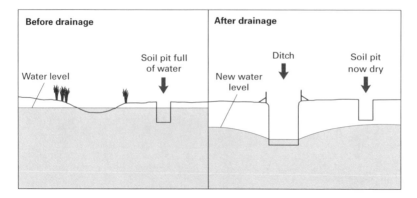

Fig. 6.3 The use of ditches to drain soils where the rate of water movement is relatively fast.

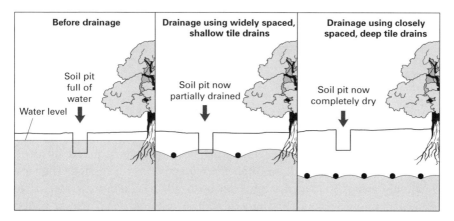

Fig. 6.4 Where the rate of water movement through the soil is very slow, subsoil drainage in addition to ditches may be needed.

used instead of ceramic pipes. They work in a similar way but have the advantage that the raw materials are cheaper (Fig. 6.4).

Mole drains are temporary structures that are created using a mole plough, which is pulled through the soil at 40–100 cm depth. The 'mole' section of the plough is a bullet-shaped implement at the base of a long metal shaft. As the plough is pulled through the soil at 2–3 m intervals, it creates a small tunnel beneath the soil surface, into which excess water drains. Mole drains are particularly effective in clay-rich soils, and despite the fact that they may need replacing every 5–12 years, they have the advantage that they are relatively cheap to install (Fig. 6.5).

When the drainage problem is due to compacted subsoil horizons, a deep plough can be used to loosen the subsoil horizons so that compacted horizons are fractured. This practice is referred to as 'subsoiling' (Fig. 6.6). One factor that determines the success of mole drains and subsoiling is

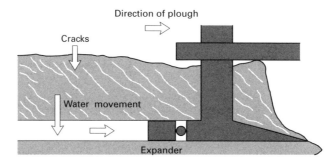

Fig. 6.5 The mole plough. (Adapted from R. E. White (1997) *Principles and Practice of Soil Science*. Blackwell Science, Oxford.)

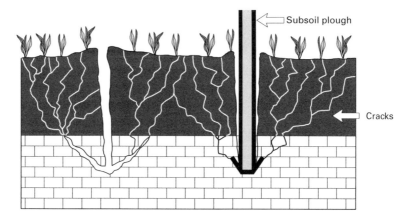

Fig. 6.6 Subsoil ploughs can be used to break up compacted soil horizons. (Adapted from R. E. White (1997) *Principles and Practice of Soil Science*. Blackwell Science, Oxford.)

soil moisture content at the time of installation. In the case of mole drains, the soil needs to be moist enough to allow the mole plough to be pulled through it cleanly without causing the soil to shatter or smear. Subsoiling, however, requires the soil to be dry enough for the impermeable layers to shatter when disrupted by the plough.

Where the drainage is very poor it may be necessary to install ditches and tile drains supplemented with mole drains or subsoiling. When mole drains and subsoiling are installed in addition to the main tile drainage system they are referred to as 'secondary treatments' (Fig. 6.7).

Drains can be laid out in a number of patterns, including 'fish bone', 'natural' and 'interception'. Figure 6.8 illustrates each of these draining patterns.

Installing drainage does have some disadvantages – chief among these is the loss of nutrients in the drainage water. This can cause pollution to

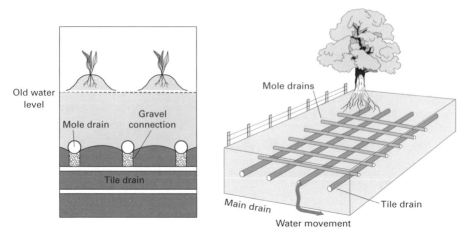

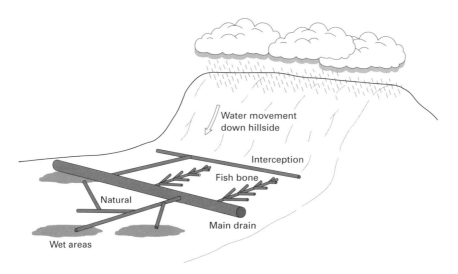

Fig. 6.7 Secondary drainage treatments can be used in soils where drainage is very poor.

Fig. 6.8 Drains can be installed using one of a number of drainage patterns. The decision which to use will be determined by local conditions.

watercourses; therefore it is important that these factors are also considered in the overall management of the land.

Problems with not enough water: storing what we have and applying extra

Under natural conditions many plants have colonized environments which are prone to water stress by developing physiological adaptations. These include reproductive cycles which are timed to coincide with rainy seasons, deep roots, waxy skins and water storage organs. With commercial crops we are limited in the type of physiological adaptations we can incorporate into the plant (for instance, imagine living on a diet of cactus). However, we can use two complementary approaches to enhance water supply. The first is to maximize the storage of water in the soil and the second is to apply extra water (irrigate). We will start by looking at how we can maximize the amount of water stored in the soil.

Maximizing the storage of available water: water conservation

Let us return to the idea of plant-available water that we discussed in Chapter 2. The amount of water stored in the soil for plant growth is termed 'plant-available water'. This is the amount of water that the soil can store between the ranges of field capacity (when the soil is very wet) and wilting point (when the soil is too dry to supply water to the plant). The amount of plant-available water that the soil can store will depend on its texture, structure and depth. It may be that, with careful management, rainfall could provide all of the crop's water requirements. The methods we can use to maximize the storage of water in the soil are:
♦ reduce evaporation losses;
♦ increase infiltration rates;
♦ use fallow periods to store water over successive years.

There is a natural assumption that a large proportion of the water falling as rain will be absorbed by the soil and later used by plants; however, this is not always the case. In densely vegetated areas, up to 40% of the rainfall may be intercepted by foliage before being evaporated back to the atmosphere, without ever reaching the soil. This process is called 'interception loss'.

Interception and evaporation losses can be particularly high in the tropical areas. Losses are further increased because of the type of heavy showers (>25 mm/h) that occur in the tropics. These have a tendency to destroy soil structure, which can lead to soil crusting (see Chapter 2), preventing the infiltration of water. In steeply sloping areas water can also be lost through 'surface run-off'. Run-off occurs when rainfall exceeds infiltration rates, and rather than collecting in pools, the water runs over the surface of the soil before being channelled straight into streams and rivers without ever penetrating the soil. In addition to water loss, run-off can also lead to soil erosion and the loss of nutrients. Figure 6.9 shows the creation of specialist cultivation measures, designed to trap moisture around seedlings.

We can reduce evaporation losses and soil crusting, and increase infiltration rates by ensuring that the surface of the soil is covered at all times with plant residues; these are called 'mulches'. Mulches protect the soil surface from the destructive impact of rainfall, help prevent run-off

Fig. 6.9 Cultivation can also be used to create specialist structures. Here soil has been used to trap rainwater in a series of hollows, increasing soil moisture and seedling survival rate.

and reduce evaporation losses. In steeply sloping areas the surface of the land can be levelled to form a terrace, which can further reduce the risk of run-off. The amount of stored water in the soil can be increased by using fallow periods and preventing evaporation losses. Preventing plant growth for a season allows the rainfall from consecutive years to be stored in the soil. Figure 6.10 summarizes some of these water conservation methods.

Adding extra water: irrigation

Despite our best efforts, in some circumstances there will always be insufficient water at the right time to meet crop demands. Water shortages may arise for a number of reasons and these are not always confined to areas with hot climates. For example, even on an overcast summer's day in the UK, a crop may transpire 25–30 tonnes of water per hectare per day. In other cases it may be that the crop is outside of its natural range of tolerance or that conservation measures such as mulches are impractical, or simply that it is a particularly dry year. In all of these situations there are few options but to irrigate.

Approximately 270 million hectares of land are now under some form of irrigation. Areas in the USA, China, India, Pakistan and Russia all have extensive irrigation schemes, and in many smaller countries – such as

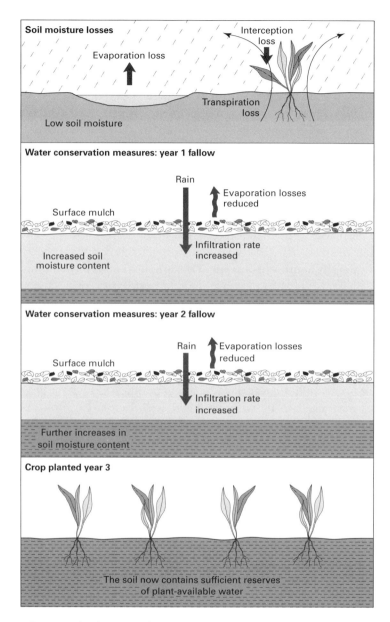

Fig. 6.10 Reserves of soil water can be increased using fallow periods and mulches. Mulches aid the infiltration of rainwater and help reduce evaporation losses. This allows water from successive years to be collected and stored in the soil.

Egypt, Israel, Sudan and Iraq – agricultural production is virtually dependent on it. H. L. Penman, a soil physicist who worked at Rothamsted Experimental Station, UK, once remarked that 'the value of successful irrigation is that it provides the water of a wet summer in the sunshine of a fine one'. There are a number of ways of applying extra water:

♦ Flooding: where water is supplied from a surface water source, such as a river or lake. This method is only suitable for relatively flat land and can be wasteful in areas with high evaporation rates.

♦ Channel flooding: is a similar method to flooding except that water flows down channels between furrows, replenishing soil water stocks.

♦ Overhead: where water is applied to the crop using sprinklers or a boom. Booms can be connected to a motor that allows it to move through the crop. In some circumstances, the boom rotates (centre pivot irrigation) and creates a large green circular patch of vegetation (you can sometimes see these when flying over irrigated areas). The disadvantages associated with overhead irrigation are the cost of equipment, the problem of irrigation water being blown off course in high winds, and the possibility that contact pesticides applied to foliage are washed off.

♦ Trickle: in soils with very weak structures, irrigation water needs to be applied carefully to prevent crusting. Trickle irrigation, as the name suggests, applies the water next to the crop using a perforated pipe. This reduces the risk of soils with weak structures forming crusts as aggregates fall apart after being hit by water drops.

♦ Subirrigation: water is supplied to the crop underground using a perforated plastic pipe. It requires very careful management to supply the crop with enough water for optimum growth while ensuring that waterlogging does not occur. In hot countries where evaporation is high, subirrigation schemes may need to be supplemented with overhead applications of water in order to control salt build-up (this is discussed in more detail later in this chapter).

Potential hazards associated with irrigation

Unfortunately, irrigation is not simply a matter of applying water to a crop. If either the amount of water applied or the quality of the water (particularly its salt concentration) is incorrect, the farmer runs the risk of not only wasting water but also damaging the soil. In some cases the damage caused by poorly managed irrigation schemes is so bad that the soil cannot be used for further agricultural production without remedial action. One example of the problems associated with irrigation is 'saline creep', which has been reported in Australia, the USA and Canada. The problem is caused by a combination of applying too much irrigation water and vegetation changes from natural forest to commercial crops; this leads to rising ground-water levels. As ground water rises it dissolves salts contained in the lower soil horizons. Eventually, the briny solution saturates the rooting zone, destroying the agriculture of the area. In Australia the problem is now being managed by extracting ground water, by pumping it into ponds where it evaporates. Salt deposits are then collected and sold, and the profits used to pay for the pumping.

In order to irrigate a crop successfully we need to know the answers to the following questions:

♦ When does the crop need water?
♦ What are the water requirements of the crop?
♦ What is the quality of the irrigation water?

Plants can only reach optimum growth if adequate supplies of water are available throughout their growth cycle. Many government agricultural services now publish tables that list the crops grown in their areas with the periods in which they are sensitive to moisture stress. Some crops have certain growth stages that are particularly sensitive to water shortages. For example, peas produce high yields as long as sufficient water is available during the times of fertilization and pod swelling. Yield suffers only when suboptimum amounts of water are available at these critical growth stages. Other examples include turnips, which require irrigation when their roots start swelling (usually between April and May), and crops such as sweet corn, which are sensitive to moisture stress throughout the whole of the growing season. Irrigation schemes need to apply enough water to satisfy crop needs at these critical times during the growing season – but not too much water, which is both wasteful and counterproductive as nutrients are leached out of the root zone. In order to supply the right amount of water we need to know:

♦ the water requirements of the crop;
♦ how much water will be supplied by natural means through rainfall and ground water;
♦ the amount of water already contained in the soil before irrigation water is applied.

The water requirements of the crop are closely related to its transpiration rate. Potential rates of transpiration are calculated using reference data from plants such as grass. These are then modified with the appropriate correction factors that adjust the transpiration rate to that of the crop. This measurement is combined with climatic data (i.e. the amount of solar radiation, wind speed and direction) and air temperature to give the maximum 'potential evapotranspiration' rate.

Water is lost from the soil as plants transpire, and by evaporation and drainage. The water status of the soil is described by expressing it in terms of the amount of water (mm) that is needed to return the soil to field capacity. This measurement is referred to as the 'soil moisture deficit' or SMD. Figure 6.11 shows the SMD for two soils. The amount of irrigation water that needs to be applied to a field is simply expressed as the depth of water that needs to be applied to the site. Therefore, an application of 25 mm means an application of 25 mm of water to the whole of the irrigated area. In many countries regional SMDs are calculated for farmers by commercial companies as part of their extension service, or by government departments. The decision whether to irrigate will depend on the type of crop being

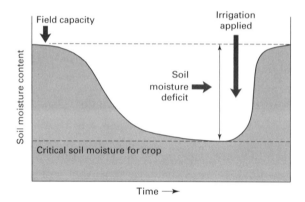

Fig. 6.11 The amount of irrigation water to apply can be related to the soil moisture deficit (SMD). Here the SMD is shown graphically over time as water is lost from the soil by drainage and evapotranspiration.

Table 6.1 Irrigation schedule for arable crops.

		Irrigation plan (mm of water at mm SMD)		
Crop	Response period	Sandy	Loam	Clays
Potatoes	June–August	25 at 25	25 at 40	25 at 40
Cereals	May–June	25 at 50	no irrigation required	

Source: ADAS/MAFF (1981) *Irrigation* (ADAS reference book 138). MAFF Publications, London.

grown: for example, some crops such as potatoes are particularly sensitive to water stress. It is recommended that in the UK, on medium-textured soils, between June and August a total of 25 mm of water is applied to potato crops when the SMD is between 25 and 40 mm. However, when cereals are grown on a similar textured soil, irrigation is usually not required. The irrigation plan for two crops is shown in Table 6.1.

In countries with generally cool climates and good quality irrigation water it is often sufficient simply to replace the water lost through evapotranspiration. However, in hot countries where the loss of water from the soil can be high, in addition to supplying water to meet the crop's needs, extra water also needs to be applied in order to control the build-up of salts in the soil. We can manage the threat posed by salt by doing two things. First, we can make sure that we only use water with salt concentrations that fall within certain safe limits. Secondly, we can add extra water (in addition to crop requirements) so that salts which have accumulated are leached out of the soil with the drainage waters. This additional water is called the 'leaching requirement'.

Table 6.2 Irrigation water classification.

Salinity	Hazard	Reading
No problem	EC	<0.75 (dS/m)
Increasing problem	EC	0.75–3.0 (dS/m)
Severe problem	EC	>3.0 (dS/m)

Table 6.3 Salt tolerance (EC of the soil when 100% crop failure is recorded).

Sensitive (4–7)	Less sensitive (8–10)	Moderately sensitive (11–15)	Moderately tolerant (15–20)	Tolerant (>20)
Strawberry	Grapefruit	Paddy rice	Spinach	Wheat
Raspberry	Orange	Cabbage	Sesbania	Sugar beet
Apricot	Apple	Broccoli	Sorghum	Cotton
Avocado	Soybean	Fig		Barley
Groundnut	Corn			Date palm
Plum				

Source: R. S. Ayers and D. W. Westcot (1976) *Water Quality for Agriculture* (Irrigation and Drainage Paper 29). Food and Agriculture Organization of the United Nations, Rome.

Unless it is treated in some way – usually by distillation – all water contains dissolved chemicals. These can be organic molecules or inorganic ions, which form salts once combined together. In most cases, the commonest ions in irrigation water are calcium (Ca^{2+}), magnesium (Mg^{2+}), sodium (Na^+), chloride (Cl^-), nitrate (NO_3^-), bicarbonates (HCO_3^-) and sulphates (SO_4^{2-}). The salt concentration of water can be determined easily by measuring how well it conducts an electric current. Water that has a high concentration of dissolved salt will also have a high electrical conductivity (EC). It is measured in millimho per cm (usually expressed as mmho/cm or dS/m). Once the EC has been measured, irrigation water can then be graded. Salinity hazard gradings are shown in Table 6.2. Even when using water with an EC of 1.15 dS/m, it has been estimated that US farmers applying 124 cm of water over one growing season will also be adding 9 tonnes of salt per hectare.

The EC values given in Table 6.2 are only meant to serve as guidelines, because the salt tolerance of different crops will vary. In some circumstances the farmer will be able to use poorer quality irrigation water if the crop has a high salt tolerance. Table 6.3 shows the salt tolerance of some common plants.

The leaching requirement is dependent on the quality of the irrigation water. It is the minimum amount of extra water that needs to be applied to control the build-up of salts in the soil. Controlling the build-up of salts

Table 6.4 FAO crop tolerance*.

Crop	0%		10%		25%		50%		Maximum
	ECe	ECw	ECe	ECw	ECe	ECw	ECe	ECw	ECe
Barley	8.0	5.3	10.0	6.7	13.0	8.7	18.0	12.0	28.0
Cotton	7.7	5.1	9.6	6.4	13.0	8.4	17.0	12.0	27.0
Corn	1.7	1.1	2.5	1.7	3.8	2.5	5.9	3.9	10.0

* ECw = EC of water; ECe = EC of soil.
Source: R. S. Ayers and D. W. Westcot (1976) *Water Quality for Agriculture* (Irrigation and Drainage Paper 29). Food and Agriculture Organization of the United Nations, Rome.

in the soil is of crucial importance when applying irrigation water in hot climates. If we do not manage irrigation correctly the soil may be made worthless for future agricultural purposes.

Let us assume that we have tested the quality of our irrigation water and it lies within acceptable limits for the crop we want to grow. Before applying irrigation water we need to calculate the leaching requirement (LR). We can calculate LR using the following calculation (taken from Irrigation and Drainage Paper 29, FAO, 1976):

$$LR = \left[\frac{\text{EC of irrigation water}}{(5 \times \text{EC of soil}) - \text{EC of irrigation water}} \right] \tag{10}$$

The amount of water to apply to the crop, including the leaching requirement, is then given by:

$$\text{Water to apply (mm)} = \left[\frac{\text{Crop water needs (mm)}}{1 - LR} \right] \tag{11}$$

Many countries have their own agricultural departments to advise farmers on when to irrigate and how much water to use. The FAO has published a set of irrigation guidelines. We can illustrate a number of the points covered in this section by looking at an irrigation plan for corn.

Example: The EC of the proposed source of irrigation water has been found to be 1.7 d/s. There is a critical period of 20 days when the crop needs to have sufficient supplies of water. The transpiration rate of the crop during this period is 5 mm/day. What irrigation plan would you recommend?

First of all, ask yourself if the irrigation water is within the acceptable limits set by the FAO. Having established that, the second thing to do is to look at the salt tolerance table published by the FAO for a wide range of crops. A small section of one of these tables including corn is shown in Table 6.4. The table shows that a soil irrigated with water that has an ECw of 1.7 d/s will tend to produce a soil with an ECe of 2.5 d/s. Although this will probably result in a 10% yield reduction, the salt concentrations are within the tolerance range for corn. The third step is to calculate the leaching requirements using the equation:

$$LR = \left[\frac{1.7}{(5 \times 2.5) - 1.7} \right] = 0.16 \tag{12}$$

Finally, the amount of water to add is the crop's evapotranspiration rate multiplied by the time we are going to irrigate (5×20 (the number of days irrigation is required) = 100) plus the LR:

$$\text{Water to apply} = \left[\frac{100}{1 - 0.16} \right] = 119 \text{ mm} \tag{13}$$

So, in order to supply crop demands and to control the build-up of salts, 119 mm of water will need to be applied across the whole site during the 20-day period.

Changing the physical state of the soil: cultivation

Of all the images the word 'farming' conjures up, that of a plough being drawn through the soil by a tractor is probably the most enduring. However, in many parts of the world farming and cultivation (sometimes referred to as tillage) are no longer synonymous. Before looking at alternatives to cultivation, let us look again at plant roots and ask 'what are we trying to achieve when we plough a soil?'

Ploughs are designed to disrupt the soil by breaking up large clods and compacted layers so that any physical barriers to root growth are removed. The aims of cultivation are:

♦ to loosen soil, including breaking any pans that may have formed, so that aeration and drainage are increased;

♦ to bury weeds (so that they do not compete with the crop for valuable nutrients) and crop residues;

♦ to create specialized structures, i.e. ridges for moisture-sensitive crops;

♦ to create a fine structure or 'tilth' in order to maximize soil–seed contact. This helps maximize the uptake of water and nutrients and allows young roots to extend and so quickly become established.

Cultivation in parts of the world such as the UK also forms an important means of residue disposal. Straw residues, for instance, will remain intact on the soil surface for some time, whereas they decompose relatively quickly when incorporated into the soil. There are several methods of cultivation. Techniques differ in their mode of soil disruption and the depth to which they penetrate the soil. They can be divided into two groups, primary and secondary treatments, depending upon how deeply they penetrate the soil. Primary cultivation techniques are used when deep ploughing is required.

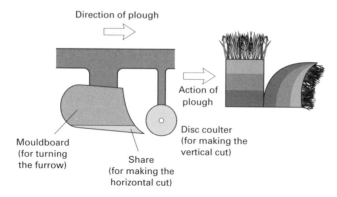

Fig. 6.12 Parts of the mouldboard plough.

Primary treatments:
$\begin{cases} \text{Mouldboard plough} \\ \text{Heavy disc} \\ \text{Chisel plough} \\ \text{Rotary cultivators} \end{cases}$

Secondary treatments:
$\begin{cases} \text{Light disc} \\ \text{Tine} \\ \text{Skim plough} \end{cases}$

The choice of cultivation techniques will depend on a combination of soil type, the crop requirements and fuel costs. The mouldboard plough (Fig. 6.12) has for many years been the most popular cultivation implement in the UK. One of the main advantages of the mouldboard plough is that it buries weeds and crop residues in addition to disrupting the soil's surface. Some heavy-textured soils may need secondary treatments following mouldboard ploughing in order to further break up large clods.

The importance of soil moisture in cultivation

Cultivation is a skilled operation that needs to be carried out at the right time, when the moisture content of the soil lies within certain limits. If we cultivate the soil when it is too dry or too wet we may damage the soil structure by forming large clods or compacting the soil into a pan. We can illustrate this point by taking a very dry soil aggregate from a heavy-textured soil. If you pick it up between your finger and thumb and squeeze it, nothing will happen. This is because when the aggregate is dry, its strength is at a maximum, so that it requires more force to break it than your fingers can exert. Now, if you add a little water, aggregate strength decreases; now we may be able to shatter the aggregate with our fingers. If we continue to add water the aggregate will smear into a paste when squeezed. Finally,

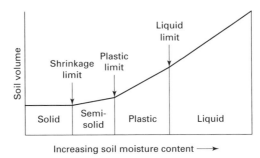

Fig. 6.13 The moisture content of a soil will affect some of its physical properties. The relationship between moisture content and solid, semi-solid, plastic and liquid soil states has important implications for cultivation and engineering.

we can add so much water that the aggregate may fall apart without any pressure to form slurry.

The decline in aggregate strength with increasing moisture content has important implications for engineering and farming. We refer to this area of soil science as 'soil mechanics'. It is primarily interested in the load-bearing properties of soils and how they differ with changing moisture content. The changing physical properties of the soil with changing moisture content are shown in Fig. 6.13.

So how can we relate soil mechanics to cultivation? In the description above, the slightly wetted aggregate tended to shatter when it was squeezed hard, but smeared into a paste when squeezed wet. A plough acts in a similar way. If a clay-rich soil is cultivated when it is too dry, it may break into large clods that will resist root growth. The other extreme is when the soil is ploughed when it is too wet so that the soil smears to form a compacted horizon called a 'plough pan'. The best time to plough the soil is when the soil is wet enough to break into small aggregates but not so wet that it smears and compacts to form a pan. When it is in this optimum state it is said to be 'friable', which basically means it is easily crumbled. This moisture content lies between two points. The drier end of the range is referred to as the 'shrinkage limit': which means that as the soil dries further there is no change in its overall volume. The other point is called the lower 'plastic limit': this is reached when the soil passes from being brittle to a substance that can be moulded (refer to Fig. 6.13). Engineers can measure it accurately whereas farmers will be able to gauge it by eye or by rubbing the soil between their fingers.

Despite their advantages, in many areas of the world the use of primary cultivation implements (such as the mouldboard plough) is not appropriate. There have been examples where these cultivation techniques have proved disastrous when exported to countries such as America and Australia. Now alternative cultivation techniques have been developed which seek to reduce soil disruption to an absolute minimum. Some of these techniques, such as 'reduced tillage', still employ some form of mechanical soil loosening, whereas others, such as 'direct drilling' or 'no-till' systems of cultivation, employ no soil disturbance. Reducing the amount of soil disturbance can have a number of advantages; these are:

♦ fuel savings;
♦ increased soil strength;
♦ reduced erosion;
♦ increased moisture conservation.

In 'no-till' systems the new crop is sown into the residues of the previous crop or into grass sods. One of the main advantages of these reduced cultivation techniques is that the soil surface is rarely exposed and this helps reduce the potential for erosion and evaporation losses. In addition to these advantages, there are also fuel savings resulting from a reduction in energy use. This is because energy expenditure is largely related to how deep the plough cuts through soil. For example, ploughing 1 hectare to a depth of 20 cm requires moving approximately 3000 tonnes of soil; ploughing to 10 cm moves half as much. However, the fuel savings with no-till are partially offset by the increased use of herbicide, because weeds cannot be controlled by burial, as happens in more traditional cultivation systems.

Research has shown that although no-till soils often have higher bulk densities, because of an overall reduction in the number of macropores, this does not seem to affect infiltration rates. This has been attributed to the fact that the pore network in no-till systems has greater continuity. For example, in no-till systems worm and root channels often extend from the surface horizons into the subsoil. Unlike in traditional cultivation systems, these channels are not disrupted by the plough and so allow transmission water to drain away quickly.

It is estimated that 65% of the crops in the USA are now under some system of reduced cultivation or no-till system. In the UK, the figure remains below 30%. The difference is primarily due to climate. In the UK, high rainfall and the need to incorporate crop residues and control weeds mean that primary cultivation techniques are likely to remain popular.

Essential points

♦ Plants require the soil to provide water, nutrients, environmental buffering and anchorage.
♦ Problems with excess water include low oxygen availability, toxin accumulation, poor nitrogen use and weed problems.
♦ Land can be drained using surface drains such as ditches, and subsurface drains such as tile drains, mole drains and subsoil ploughing.
♦ The aim of these schemes is usually to allow the field to reach field capacity within 48 h or, if the problem is due to a high water table, to lower the water table so that it lies 2 m below the soil surface.

♦ Drains can be arranged using a variety of patterns, which include fish bone, natural and interception.

♦ When the problem is due to too little water we can maximize the effectiveness of the available water by conservation measures, like reducing evaporation, increasing infiltration and storing water over consecutive years using fallow periods.

♦ If these measures are ineffective, irrigation can be used. Water can be applied using a variety of different methods. Some methods are simple, such as flooding; others require substantial investment in water storage facilities and application machinery.

♦ If managed incorrectly, irrigation can ruin the fertility of the soil. Before irrigation water is applied the quantity of salt in the water must be known. From this figure we can calculate the leaching requirement. This helps prevent the build-up of salt.

♦ Cultivation can help create optimum rooting conditions and control weeds.

♦ Cultivation implements can be regarded as primary or secondary treatments, depending upon how deeply they penetrate the soil surface.

♦ It is essential that cultivation is carried out when the soil is at the optimum moisture content. For most soils this lies between the shrinkage limit and the plastic limit.

♦ In some climates primary methods of cultivation are not recommended. In areas that have potentially high rates of erosion minimum till or no-till is practised.

2   How can we optimize the chemical condition of the soil for plant growth?

Providing the plant with the right chemical environment

In addition to providing the best physical conditions for plant growth, we also need to optimize the chemical environment by preventing the build-up of toxins and by supplying nutrients. In agricultural terms the most important way we can do this is by controlling acidity through the application of lime, and applying chemical fertilizers to supply nutrients.

How can we control acidity by applying lime?

Even under natural conditions soils have a tendency to become acidic. This is mainly because of rainfall, microbial respiration and nutrient uptake by crops. Even in non-polluted areas rainfall is slightly acidic, so after prolonged leaching the pH of the soil and rainfall come to equilibrium at a pH between 5 and 5.6. In addition, microbial respiration leads to the emission

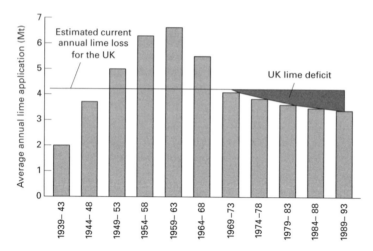

Fig. 6.14 Average annual lime application rates for the UK, averaged over 5-year periods. (From K. Goulding and B. Annis (1998) *Lime, Liming and the Management of Soil Acidity*, Proceedings No. 410. The Fertiliser Society, York.)

of carbon dioxide, which produces a weak acid once dissolved in the soil water. However, in agricultural soils the rate of acidification is often greater because of the effects of some fertilizers (particularly those containing ammonium) and the removal of base cations (which counteract acidity) from the soil after the plants have been harvested.

Soil pH also has an important effect on the availability of several plant nutrients. The optimum pH for maximum nutrient availability varies depending on the nutrient; generally 6.5–7.0 is the recommended pH range for arable soils (see Fig. 3.13). If the pH of the soil is outside this optimum range we must employ measures to change it, otherwise the crop will not reach its full potential.

We can change soil pH by applying chemicals: when the pH is too high we can add ferrous sulphate or sulphur; if it is too low we add lime. In most cases the soil will have a tendency to become acidic. It is estimated that two-thirds of all the arable soils in the UK will need lime at some time. Lime applications in the UK have been declining since the late 1960s, so that now many soils in the UK, especially upland areas, have become acidic (Fig. 6.14). The situation is particularly serious in grassland soils, where recent surveys have shown both a 20% decrease in the land area limed, and a 20% decrease in the application rate. Soil pH values below pH 4.0 may affect roots directly, but even in less acidic soils high concentrations of acidity may cause leaves and stems to show yellow mottling, stunted growth and purple veining. Some crops are more sensitive than others to high concentrations of acidity. Table 6.5 shows the acid tolerance for several crops.

Table 6.5 Plants grouped according to their tolerance to acidity.

Very sensitive	Sensitive	Moderate tolerance	Very tolerant
Barley	Corn	Oats	Blueberry
Sugar beet	Wheat	Tobacco	Cranberry
Cabbage	Soybean	Potato	
Onion			

Source: California Fertilizer Association, Soil Improvement Committee, G. R. Hawkes *et al.* (1985) *Western Fertilizer Handbook*. The Interstate Printers & Publishers, Danville, IL.

A high degree of acidity causes problems for several reasons. When soil pH falls below 5.5 aluminium ions start to dominate exchange sites (see Chapter 3). High concentrations of Al^{3+} interfere with cell division and root elongation. One sign of Al^{3+} toxicities is the formation of brown stubby roots. In addition, high concentrations of acidity render other nutrients such as phosphorus insoluble and potentially harmful metals soluble.

The main way we can control acidity is by applying lime. The active ingredient can be calcium carbonate or calcite ($CaCO_3$), calcium oxide (CaO) or magnesium carbonate or dolomite ($MgCO_3$). Lime reduces soil acidity because it reduces the concentration of H^+ in solution by the following (simplified) reaction:

$$2H^+ + CaCO_3 \rightarrow Ca^{2+} + CO_2 + H_2O \qquad (14)$$

As H^+ ions are taken out of soil solution the proportion of OH^- ions to H^+ ions increases and Ca^{2+} ions bind to exchange sites, reducing levels of exchangeable acidity. The harmful effects of Al^{3+} are reduced when it combines with OH^- to form the mineral gibbsite ($Al(OH)_3$), which is insoluble. Lime therefore reduces H^+ ion concentration, increases the concentration of Ca^{2+}, and takes Al^{3+} out of soil solution by converting it to a relatively insoluble substance.

Liming materials and application rates

The term 'agricultural lime' covers a whole range of material, from crushed seashells to basic slag from blast furnaces. Now most liming material is derived either from crushed or burnt limestone or chalk deposits. Limestone varies in its purity, ranging from 75 to 99% pure calcite or dolomite. One problem with crushed limestone is that it can be slow-acting, but we can speed up the reaction by heating it to produce quicklime (CaO). Sometimes quicklime is mixed with water to produce hydrated lime ($Ca(OH)_2$).

Table 6.6 Lime recommendation for an arable soil to a depth of 20 cm.

Soil type	Recommended pH	Target pH	Lime factor for 1 pH unit of change (t/ha)
Sands	6.5	6.7	6
Light	6.5	6.7	7
Medium and clay	6.5	6.7	8
Organic	6.2	6.4	10

Source: K. Goulding and B. Annis (1998) *Lime, Liming and the Management of Soil Acidity*, Proceedings No. 410. The Fertiliser Society, York.

Liming materials are evaluated by comparing their ability to neutralize acidity relative to calcium oxide. This is termed their 'neutralizing value' or NV. It is usually expressed as a percentage; for example, ground limestone is graded as NV50 whereas hydrated lime is NV70. The system used in the UK calculates how much lime to apply using a combination of:
♦ soil type;
♦ soil pH;
♦ target pH (the optimum pH required for the crop).
Soils that have predominantly clay textures will require more lime to neutralize acidity than sandy soils, because of their higher buffering capacities (remember the hot-water tank analogy in Chapter 3). The 'lime requirement' is the amount of lime (t/ha) that is needed to raise the soil pH to the target pH of the crop over a depth of 20 cm. Table 6.6 lists the lime requirements for arable crops on a range of soil types.
    Lime requirement is then calculated as follows:

(target pH − soil pH) × lime factor = lime requirement   (15)

Example: Before a crop of sugar beet is planted we measure the pH of the soil. We find that it is 5.0 and below the critical value of 5.9. The field therefore needs to be limed. The soil has a sandy texture. With reference to Table 6.6, the lime requirement can then be calculated:

(target pH − soil pH) × lime factor = lime requirement   (16)
(5.9 − 5.0) × 6 = 5.4 (t/ha)

In the USA, the amount of lime to apply is given with reference to six soil types and five pH ranges (as shown in Table 6.7).

Supplying nutrients: fertilizers

Under natural conditions nutrients are ultimately supplied from rock weathering and atmospheric sources. They are then continuously recycled

Table 6.7 Amount of limestone needed to change the soil reaction.

| pH required | kg of limestone per hectare | | | |
	Sand	Sandy loam	Silt loam	Clay loam
4.0–6.5	2 914	5 604	9 415	11 208
4.5–6.5	2 466	4 708	7 846	9 415
5.0–6.5	2 018	3 811	6 277	7 398
5.5–6.5	1 345	2 914	4 483	5 156
6.0–6.5	673	1 569	2 466	2 690

Source: Adapted from California Fertilizer Association, Soil Improvement Committee, G. R. Hawkes *et al.* (1985) *Western Fertilizer Handbook.* The Interstate Printers & Publishers, Danville, IL.

through the soil (see Chapter 4). Some soils, such as light sands, have naturally low nutrient reserves. In other cases, the soil may initially have sufficient stocks of nutrients to meet crop demands but gradually become depleted as successive crops are grown. Fertilizers are required when nutrient supplies from all other sources, such as SOM, are insufficient to meet crop demand.

Many crops respond well to fertilizer additions. This means that in most cases fertilizers are a sound investment. The aim of any fertilizer recommendation scheme is to apply fertilizer in such a way that it increases crop growth to its economic maximum, but at the same time ensuring that pollution to the environment is kept to a minimum. In order to achieve these objectives we need to apply the right quantities of fertilizer at the right time.

Although some crops have slightly different nutrient demands (especially for micronutrients), generally nitrogen, phosphorus, potassium, sulphur and magnesium are required in relatively large quantities by all plants. Some crops, for instance, may require extra applications of trace nutrients. If too little fertilizer is applied the crop will not reach its full potential; if too much is applied, in addition to wasting money, we run the risk of causing declines in the quality of the produce and increasing the chance of other physiological problems, such as the collapse of the stem of the plant (known as 'lodging').

The amount of fertilizer to apply can be determined by measuring the following:
♦ the nutrient requirement of the crop measured at the expected average yield;
♦ the soil's ability to supply nutrients;
♦ nutrient losses from the soil.

Over the years agricultural research stations throughout the world have attempted to optimize fertilizer application rates so that farmers achieve the maximum economic returns. In fertilizer trials plants are grown with

Fig. 6.15 Fertilizer trials help determine the correct timing and application rate for fertilizer. Shown here is the longest-running field trial in the world, located at IACR-Rothamsted. (Courtesy of the Photographic Department, IACR–Rothamsted.)

or without varying amounts of fertilizer. This allows the application rate and the timing of applications to be assessed. Trials such as these are often used to assess the effects of other treatments on crop growth, such as organic matter and lime additions. The photograph in Fig. 6.15 shows one such trial; note how the different treatments have produced colour differences between the fields.

These trials have shown that fertilizer additions do cause dramatic increases in crop growth and quality, but as more fertilizer is applied, yield increases get smaller and smaller, until the cost of the extra fertilizer is greater than any further increases in yield. This observation is called the law of 'diminishing returns'. When very high levels of fertilizer are applied, toxic effects may cause declines in crop growth. A typical growth response curve is shown in Fig. 6.16.

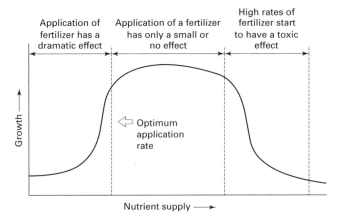

Fig. 6.16 Crop growth in response to additions of fertilizer.

Fertilizer trials have produced a broad set of data on which general management advice can be given to farmers. In the UK the Ministry of Agriculture, Fisheries and Food (MAFF) has published a set of guidelines that farmers can use to optimize fertilizer application for a number of different crops grown under a range of environmental conditions. In order to illustrate how a recommendation system works, we will look at the scheme developed by MAFF for calculating nitrogen, phosphorus and potassium application rates.

Nitrogen

Most crops require relatively large amounts of nitrogen. Nitrogen has a greater effect on crop yield and quality than any other nutrient. Nitrogen shortages can be detected by leaf analysis and checking for symptoms such as pale leaves and fast-maturing plants. At the other extreme, when too much nitrogen is applied plants tend to develop deep green bushy foliage and have slow maturing times. In the MAFF system the optimum application of nitrogen can be calculated using a combination of the following:

♦ rainfall;
♦ soil type;
♦ last crop grown.

The amount of rainfall an area receives will determine the potential nutrient losses through leaching. Soil type, particularly its texture, will determine several factors, such as soil nutrient reserves, how efficiently the crop will use the fertilizer, and how effectively nutrient residues from the previous crops are returned to the soil. For example, in sandy soils an average crop will use 70% of the applied fertilizer, whereas in heavier-textured soils, such as silts and clays, only 60% of the applied fertilizer will be used. The previous crop will determine both the amount and the type of residues left in the soil. Residues that are easily mineralized (i.e. those which have small C to N ratios) will return more nitrogen to the soil in the next growing season than residues with larger C to N ratios. This extra nitrogen needs to be taken into account when calculating how much fertilizer needs to be applied in the next growing season.

One of the critical factors in applying the right amount of nitrogen to the soil is being able to predict how much of the crop's nitrogen requirement will be met from the soil's own store of nitrogen. This factor has been termed 'soil nitrogen supply' or SNS. It will depend largely on the texture of the soil and the type of residues that have been returned to the soil from the previous crop. However, not all the nitrogen supplied by the soil will be available to the plant because a proportion of the SNS will be lost by leaching, denitrification and ammonium volatilization. These losses need to be taken account of before calculating how much

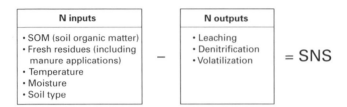

Fig. 6.17 The amount of fertilizer nitrogen to apply can be calculated by considering the soil nitrogen supply (SNS). This is the input of nitrogen minus the outputs. Estimated crop demand, minus SNS, will be equal to the amount of fertilizer nitrogen required.

nitrogen to supply to the crop. The calculation of the SNS is shown schematically in Fig. 6.17.

In the UK, nitrogen losses are estimated by dividing the country into three areas depending upon rainfall: those areas with less than 600 mm, those with 600–700 mm and those over 700 mm per year. Fertilizer recommendations can then be given on the basis of soil type and rainfall area, with this information being used to calculate an index value. The index value is then used to determine how much nitrogen to apply. Table 6.8 shows how an index value is calculated for a range of soil types in an area with moderate rainfall (600–700 mm/yr) with reference to the last crop grown. Table 6.9 shows how this index can be used to calculate the amount of nitrogen to apply to winter wheat.

Phosphorus and potassium

The aim of applying the right amounts of phosphorus and potassium is the same as for applying nitrogen; that is, to apply the right amount of fertilizer at the right time. Plants can take up phosphorus as $H_2PO_4^-$ or as HPO_4^{2-}. The availability of phosphorus is closely linked to soil pH. Deficiencies are usually detected by stunted growth, lack of vigour, poor grain and seed development, and purplish or bluish coloration in leaves and stems. Plant response to phosphorus will be determined to some extent by species. Potassium deficiencies can be detected by withered leaf tips, lower drought resistance, weak stalks, and yellowing of leaves. Its availability is strongly influenced by soil texture, specifically clay content and the type of clay minerals present.

Index values for phosphorus and potassium are calculated directly by measuring the concentration of extractable phosphorus and potassium in the soil. On the basis of this measurement soils can be divided into nine index classes. For most agricultural crops, the aim is to keep phosphorus and potassium concentrations at around 16–25 and 120–180 mg/L of soil respectively, which corresponds to an index value of 2. Extra additions of fertilizer to build up phosphorus reserves can take a number of years to become effective. Potassium, on the other hand, tends to be more mobile.

Table 6.8 Soil nitrogen supply (SNS) indices for moderate rainfall areas (600–700 mm annual rainfall, or 150–250 mm excess winter rainfall), based on the last crop grown.

SNS index:	0	1	2	3	4	5	6
SNS (kg/ha N):	<60	61–80	81–100	101–120	121–160	161–240	240>
Light sands or shallow soils over sandstone	Cereals, potatoes, peas, beans, oilseed rape, forage crops (cut), low and medium N vegetables	Sugar beet, high N vegetables, rotational set-aside					
Medium soils or shallow soils (not over sandstone)		Cereals, sugar beet, forage crops (cut), low N vegetables	Peas, beans, potatoes, oilseed rape, medium N vegetables	High N vegetables, rotational set-aside			
Deep clay soils		Cereals, sugar beet, forage crops (cut)	Peas, beans, low N vegetables	Oilseed rape, potatoes, medium N vegetables	High N vegetables, rotational set-aside		
Deep fertile silty soils		Cereals, sugar beet, low N vegetables, forage crops (cut)		Oilseed rape, potatoes, medium N vegetables	Peas, beans, high N vegetables, rotational set-aside		
Organic soils							
Peat soils						All crops	

SNS = SMN (0–90 cm soil depth) + crop N + estimate of mineralizable N.
Source: MAFF (2000) *Fertiliser Recommendations for Arable and Horticultural Crops* (RB 209), 7th edn. MAFF Publications, London.

Table 6.9 Wheat (autumn and early winter sown: SNS index used to calculate nitrogen required.

	SNS index (kg/ha)						
	0	**1**	**2**	**3**	**4**	**5**	**6**
Light sand soils	160	130	100	70	40	0–40	0
Shallow soils over chalk		240	200	160	110	40–80	0–40
Medium and deep clay soils, shallow soils over rock (not chalk)		220	180	150	100	40–80	0–40
Deep fertile silty soils		180	150	120	80	40–80	0–40
Organic soils				120	80	40–80	0–40
Peaty soils							0–60

Source: MAFF (2000) *Fertiliser Recommendations for Arable and Horticultural Crops* (RB 209), 7th edn. MAFF Publications, London.

Table 6.10 Target soil indices for phosphorus and potassium.

Crop	Soil phosphorus (mg/L)	Soil potassium (mg/L)
Arable, forage, grassland	16–58	120–180
Vegetables	26–45	181–240

Source: MAFF (2000) *Fertiliser Recommendations for Arable and Horticultural Crops* (RB 209), 7th edn. MAFF Publications, London.

Table 6.10 shows the recommended soil concentration of potassium and phosphorus for arable crops.

Once the desired index concentration has been reached, it is maintained by applying the same quantity of phosphorus and potassium that the crop removes each growing season. This is determined by look-up tables that list a range of crops with the average off-take concentrations of phosphorus and potassium at harvest. Table 6.11 shows the phosphorus and potassium concentrations of several crops.

In addition to predictive fertilizer recommendations such as the MAFF scheme, which is based on soil and climate factors, the crop can also be tested for nutritional deficiencies throughout its growth cycle. Values like those shown in Table 6.12 for broccoli and cabbage, give the recommended nutrient concentration for plant tissue in a range of crops. If the nutrient concentration of the crop is below these limits the farmer needs to apply fertilizer.

Table 6.11 Phosphate and potash in crop material.

Crop	Section	Phosphate (P$_2$O$_5$) (kg/t of fresh material)	Potash (K$_2$O)
Cereals	Grain only (all cereals)	7.8	5.6
	Grain and straw:		
	winter wheat/barley	8.6	11.8
	spring wheat/barley	8.8	13.7
	winter/spring oats	8.8	17.3
Oilseed rape	Seed only	14.0	11.0
	Seed and straw	15.1	17.5
Peas	Dried	8.8	10.0
	Vining	1.7	3.2
Field beans		11.0	12.0
Potatoes		1.0	5.8
Sugar beet	Roots only	0.8	1.7
	Roots and tops	1.9	7.9
Grass	Fresh grass (15–20% DM)*	1.4	4.8
	Silage (25% DM)*	1.7	6.0
	Silage (30% DM)*	2.1	7.2
	Hay (86% DM)*	5.9	18.0

*DM = dry matter.
Source: MAFF (2000) *Fertiliser Recommendations for Arable and Horticultural Crops* (RB 209), 7th edn. MAFF Publications, London.

Table 6.12 Plant tissue analysis guide for crops.

Crop	Sampling time	Nutrient	Nutrient level (ppm) Deficient	Sufficient
Broccoli	Midgrowth	N	7 000	10 000
		P	2 500	5 000
		K	3	5
Cabbage at heading		N	5 000	9 000
		P	2 500	3 500
		K	2	4

Source: Adapted from California Fertilizer Association, Soil Improvement Committee, G. R. Hawkes *et al.* (1985) *Western Fertilizer Handbook*. The Interstate Printers & Publishers, Danville, IL.

Types of fertilizer and the timing of applications

Fertilizers can be applied in several physical forms. In order to make an informed decision about which one to choose we need to know the following:

♦ the nutrient concentration of the fertilizer;
♦ the chemical form of each nutrient;
♦ the physical nature of the fertilizer, i.e. liquid or solid;
♦ the cost of the fertilizer in £/kg nutrient (not £/kg of fertilizer);
♦ the speed of effect (fast or slow acting).

Fertilizer should be applied when the crop risks running short of essential nutrients. This will depend to some extent on the crop, but generally crops are vulnerable early in the growing season when there is a combination of rapid growth coupled with an underdeveloped root system. Nitrogen can be applied in a number of forms such as urea, ammonium nitrate, ammonium sulphate, and calcium ammonium nitrate. Phosphate can be applied in forms that are soluble such as Superphosphate and Triplephosphate, or in other less soluble forms, such as rock phosphate, which tend to act over a longer period of time. Phosphorus is notoriously difficult to keep in a plant-available form, particularly if the optimum pH of the crop falls outside the narrow limits of phosphorus solubility. Of the phosphorus applied, typically only 10–20% is absorbed by the plant during the first growing season. The remainder will be converted into insoluble products, which only slowly become available.

Fertilizers and the environment: how can we minimize the pollution risk?

In 1991 the EU passed a directive that required member states to reduce nitrate pollution in surface and ground waters when concentrations exceeded 50 mg/l. In response to this, 68 nitrate-vulnerable zones were set up throughout England and Wales in 1998, covering a total area of 600 000 ha. Farmers in these areas were required to follow guidelines designed to reduce soil nitrate concentrations. The main points were:

♦ a closed period for the application of fertilizer in autumn and winter;
♦ a closed period for the application of slurry, sewage sludge and other manures on shallow and sandy soils;
♦ a limit of 250 kg N/ha to grassland and 170 kg N/ha to arable land to be applied as farm-based manure over the whole farm;
♦ a limit of 250 kg N/ha of farm-based manure to individual fields;
♦ fertilizers and manure not to be applied to steeply sloping land or to waterlogged, frozen or snow-covered fields;
♦ farmers must apply fertilizers using recommendation systems such as those developed by MAFF;

♦ fertilizers and manure not to be applied within 10 m of watercourses;

♦ a demand that farms have sufficient storage capacities for manure.

The EU is now looking at similar measures for reducing ammonium emissions from intensely reared livestock. Some pig and poultry production is already being governed by legislation.

Losses of phosphorus from agricultural land can be partially hazardous. Streams and lakes enriched with phosphorus can suffer algae blooms, which can then cause oxygen shortages that alter the biology of the ecosystem. This process is called 'eutrophication'. Phosphorus losses can be reduced by:

♦ reducing erosion losses (phosphorus is often absorbed onto soil particles);

♦ maintaining phosphorus concentration in the soil by using the index scheme so that the build-up of unnecessarily high concentrations of phosphorus is avoided.

Essential points

♦ In order to provide the right chemical environment for plant growth, the build-up of acidity must be prevented and nutrients supplied at the right time and in the right quantities.

♦ Most soils, even under natural conditions, become acidic over time. However, many farming practices tend to increase this by removing base cations and through the application of fertilizers.

♦ Soil acidity can be controlled by adding liming materials. The amount of lime to add will be determined by: soil type, soil pH and the target pH of the crop.

♦ Agricultural soils will face a net loss in nutrients with each harvest unless the nutrients are replaced.

♦ The aim of applying fertilizer is to ensure that crop growth reaches its economic maximum and that waste and pollution from excess fertilizer is kept to a minimum.

♦ Agricultural research stations around the world have conducted fertilizer trials. The aim of these trials is to optimize the timing and the application rate of fertilizer.

♦ In the UK, application rates of nitrogen are determined by using a combination of rainfall data, soil type and information on the last crop grown. Phosphorus and potassium application rates are determined by measuring the concentration of phosphorus and potassium in soil and then referring to index values and replacing the amount removed by the crop.

♦ If fertilizers are applied at the wrong time and at the wrong application rate the farmer risks wasting money and polluting the environment.

3   How can we optimize the biological condition of the soil for plant growth?

Using natural biological cycles to maintain soil fertility

The mass production of chemical fertilizers has been a relatively recent development in agriculture: yet farming has been practised in one form or another for thousands of years. This raises the question, 'How did early farmers grow crops without major inputs?'

Part of the answer lies in the management of biological factors. One of the earliest examples is 'shifting cultivation' or 'slash and burn', which is still practised in some parts of the world. It is one of the most basic of all cultivation practices, which manages soil fertility by using natural cycles of regeneration. The cycle starts with forest or savannah clearance using fire. The crop is then planted directly into the charred remains of the native vegetation. This acts as surface mulch and helps supply some nutrients to the crop. Within 1 to 5 years, weed problems and declining fertility force the farmer to move on to another area where the process is repeated, as shown in Fig. 6.18. Shifting cultivation works well when there are long regeneration periods (10–20 years). However, the system breaks down when regeneration times are reduced and the secondary regrowth vegetation is prematurely cleared before the nutrient balance of the soil is restored. Population pressures have led to decreasing fallow periods, so that in many areas where shifting cultivation is still practised soil fertility is now in decline.

Early farmers in Western Europe faced a similar problem centuries ago. Population pressures forced farmers to examine new ways of managing the land. Rotational cropping systems were developed, whereby different crops were alternated with livestock in fallow years (mixed farming). This allowed both continuous cultivation and the maintenance of soil fertility.

If we compare crop rotations with shifting cultivation we see that as the forest regenerates, vegetation increases and tree roots help stabilize soil structure. If leguminous plants are present, nitrogen will be fixed from the atmosphere and returned to the soil as leaves and roots die. Soil organic matter will also increase over very long periods of regeneration. In addition, crop pests and weeds that may have taken hold are slowly eradicated by a lack of food and increased competition from other organisms as the forest slowly regenerates.

The aim of crop rotation is to recreate some of these effects using careful management, so that nutrients extracted by the crop are returned before a new crop is sown, and good soil structure is maintained. Good rotational systems should allow the farmer to maintain both soil fertility and income.

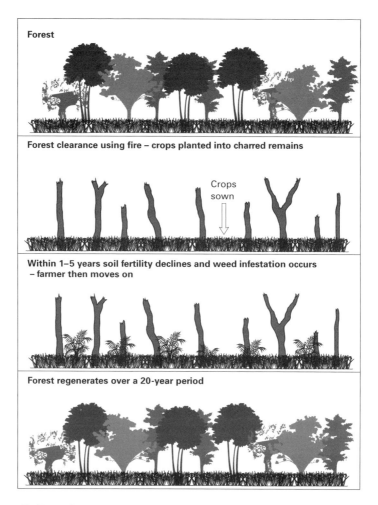

Fig. 6.18 Shifting cultivation involves the clearance of native forest and the cultivation of the soil for a number of years, before declining soil fertility and weed problems eventually force the farmer to move on to a new area. The cycle is then repeated.

One such rotational system developed in England was the 'Norfolk four-course rotation' and is shown in Fig. 6.19.

Organic manure: benefits and potential problems

Soil organic matter plays a crucial role in the maintenance of soil fertility. The main benefit of organic manures is their ability to supply nutrients, particularly nitrogen. However, their beneficial effects are not solely attributed to nutrient supply because there is evidence that organic residues can also improve the physical properties of soils. There are several organic residues that can be applied to the soil, such as:

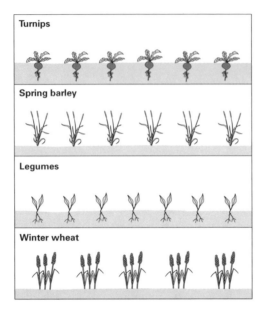

Fig. 6.19 The Norfolk four-course rotation enabled farmers to manage the fertility of their soil by varying the type of farming they practised.

♦ FYM (farmyard manure): animal dung and straw;
♦ poultry droppings;
♦ slurry: FYM in a suspension of urine;
♦ sewage sludge (raw or composted);
♦ various organic materials including wood ash, sawdust, wood chippings or seaweed.

If used correctly, organic manures can recycle waste and provide valuable fertilizer. However, if used unwisely they can have disastrous environmental consequences. In the UK, MAFF has recommended that no more than 250 kg/ha of organic manure is applied to agricultural land. It is believed that application rates in excess of this figure may cause unacceptably high rates of nitrate and phosphate leaching, coupled with high ammonia losses through volatilization. Where manure is used in conjunction with inorganic fertilizers, the nutrient content of the manure must be taken into account before the fertilizer is applied. Guidelines include tables of generalized nitrogen, phosphorus and potassium concentrations of some commonly used organic manures (Table 6.13).

Plants can only use the nutrients contained within manure once it has been mineralized by soil organisms. Different types of manure will be mineralized at different rates. For example, in poultry manure 40–60% of the N is termed readily available, whereas in FYM the figure is more likely to be 10–25%. We can use these figures to calculate how much nitrogen, phosphorus and potassium will be supplied by the manure: this can then be compared with the nutrient requirements of the crop. If there is a shortfall

Table 6.13 Typical N, P and K content of some animal manures.

Manure type	Dry matter (%)	Nutrients (kg/t)		
		N	P_2O_5	K_2O
Cattle manure	25	6.0	3.5	8.0
Pig manure	25	7.0	7.0	5.0
Sheep manure	25	6.0	2.0	3.0

Source: MAFF (2000) *Fertiliser Recommendations for Arable and Horticultural Crops* (RB 209), 7th edn. MAFF Publications, London.

in the nutrient supply, fertilizer can be applied. Careful planning enables growers to reduce the risk of over-fertilizing the soil and causing unnecessary environmental damage.

There is now a reawakening of interest in the role of biological factors in preserving and increasing soil fertility. 'Organic farming' systems view the soil as a closed system, in which outputs in the form of nutrient losses and crop removal must be balanced by inputs. Inputs such as composts, and green manures such as legumes, are promoted, while the application of synthetic fertilizers and pesticides is not permitted. Weed control is managed by mechanical and thermal means, rather than through the application of chemicals.

Many non-organic farmers argue that some of the doctrines now seen as 'organic' are simply good farming practices that have been carried out in conventional farming systems for many years. However, the growth in public interest in organic produce has stimulated research into new sources of material that can be used as fertilizers. Possibilities such as using composted rubbish from cities and towns are now being investigated.

Essential points

♦ Early farmers were expert at using biological cycles to manage the fertility of their farming systems. Fire often played an important role in the form of slash-and-burn agriculture, which was practised in many parts of the world.
♦ Slash-and-burn methods work well when long regeneration times are used. This allows the nutrient balance of the soil to be restored. Population pressure in many countries has meant that where slash-and-burn methods are still used, they are often characterized by shorter regeneration times, and this has led to declines in soil fertility.
♦ Crop rotations can maintain the fertility of the soil as well as keeping it in production. Crop rotations usually involve grain crops alternating with legumes, livestock and root crops.

♦ Organic manure can play an important role as a fertilizer and as a soil physical conditioner. However, if applied incorrectly, such as at the wrong time of the year or at the wrong application rates, it can pose a serious pollution threat.

Chapter Summary

The role of the soil scientist in agriculture is to understand and manage the factors that determine soil fertility. Plants require the soil to provide nutrients, water, environmental buffering and anchorage. When the soil is too wet, the supply of oxygen to roots is reduced. Anaerobic conditions can also lead to the build-up of harmful toxins. In order to reduce the water content of the soil, drainage can be installed. Drainage schemes can take the form of ditches, tile drains, mole drains and subsoiling. When the problem is too little water, water conservation measures such as mulches and fallow periods can be used to maximize the amount of water held within the soil. In situations where conservation techniques are insufficient, irrigation measures can be installed. Before applying water to the soil it is important that its quality is assessed; this is done by measuring its salt concentration. We can use this information to calculate the leaching requirement. The physical state of the soil can be altered using cultivation. However, care must be taken to cultivate when the moisture content of the soil lies between its shrinkage and its plastic limit.

We can create the optimum chemical conditions for plant growth by managing soil acidity and supplying nutrients. Acidity is managed by applying lime, which increases the pH of the soil. Nutrients can be supplied using fertilizer. Fertilizer must be applied in a way that provides maximum economic returns for the farmer but which does not cause pollution to the wider environment. Fertilizer trials are used to determine when is the best time to apply fertilizer and at what application rate. Early farmers had to rely on biological factors to restore the fertility of the soil. Slash-and-burn agriculture is one of the first examples of how man has sought to manage the environment. Increasing population pressures have led to shorter regeneration periods and declining fertilities. In response to these problems farmers have developed ways of alternating their farming practices by using crop rotations. Crop rotations manage the fertility of the soil without the need for long fallow periods. Organic manure can also form a useful soil amendment by supplying nutrients and improving the physical condition of the soil. Like fertilizers, they need to be applied at the right time and in the right quantities in order to prevent pollution.

For the most part, the history of soil science has been closely linked with agricultural developments. Now there is a whole set of new challenges in the form of soil pollution. Soil contamination, remediation and erosion are the topics of our final chapter.

7 Soil Contamination and Erosion

Introduction

Progress and technical innovation tend to go in fits and starts. Historically, some periods are characterized by gradual developments and others by rapid change. Today, we may be just about to embark on a 'biotech' revolution based upon the ability to create new life forms by genetic manipulation. Around 10 000 years ago another enormous change was under way as many hunter–gatherer tribes started the transition from a nomadic way of life to farming. However, as societies grow and become more complicated, it is sometimes easy to forget about the importance of the soil. It was this oversight that sealed the fate of Mesopotamia.

Mesopotamia is the area of land that lies between the Tigris and Euphrates rivers (now part of Iraq). Originally the area was swampy and agriculturally unproductive, but with growing ingenuity the people of the region (called Sumerians) slowly drained and irrigated the area, turning it into highly fertile agricultural land. The farmers were known locally as 'the men of ditches and dykes'. This highly successful agricultural system went on to provide the foundations of a great ancient civilization.

Just as many of our leaders talk confidently about finding solutions to our environmental problems, so the residents of Mesopotamia probably felt equally sure about their ability to manage the natural world. However, their confidence, like ours, was misplaced. As their culture thrived, the agricultural foundations on which everything else depended were gradually being chipped away. Today, the area is a barren wasteland riddled with saline soils. So, what happened?

The demise of Mesopotamian society started with what at the time must have looked like a minor change to the environment. Forests in the upper reaches of the Euphrates were felled, causing soil from the upper slopes to be washed into the river. As a soil-loaded Euphrates flowed through the low-lying areas of Mesopotamia it re-deposited its sediment: this slowly raised the level of the riverbed. This in turn caused the water table to rise which, coupled with faulty irrigation, led to a slow salinization of the soils. Pollen analysis from the area shows a gradual shift from wheat production to barley, which is more tolerant of saline conditions.

Eventually, the soils of the region became so degraded that they were unable to support even salt-tolerant crops. When the agriculture of the region finally collapsed, it took the rest of the society with it.

In the past one of the easiest ways to solve the problem of soil degradation was simply to move on and colonize a new area of land. Some scholars believe this was a major driving force behind the expansion of the Roman Empire. However, urbanization and population growth have meant that this solution is no longer an option. In order to preserve the agricultural viability on which most of the world's societies depend, we need to protect the quality of our soil and view it as a non-renewable resource. This chapter will look at two major soil problems, and at ways in which we can monitor the 'health' of the soil, by answering the following questions:

1   **What do we mean when we say a soil is either contaminated or polluted?**
     How do soils become contaminated?
     What are the major groups of inorganic contaminants?
     The world's shortest course in organic chemistry . . .
     What are the main types of organic contaminants in soils?
     What can happen to organic contaminants once in the soil?
     How can we assess whether or not a soil is polluted?
     How can we remediate a polluted site?
     What can we do if remediation is not possible?
     How can we prevent soil pollution?

2   **What is soil erosion?**
     Soil erosion: the nature and scale of the problem
     Soil loss: the mechanisms of erosion
     How can we assess the erosion risk?
     How can we prevent soil erosion?

3   **How can we measure the 'quality' of the soil?**
     Measuring 'soil quality'

1   What do we mean when we say a soil is either contaminated or polluted?

It is important to understand the difference between 'contamination' and 'pollution'. Soils are 'contaminated' when they have elevated concentrations of chemicals (usually as a result of human activity) compared with soils that are regarded as being in pristine condition. Contamination becomes 'pollution' once these elevated concentrations begin to have an adverse

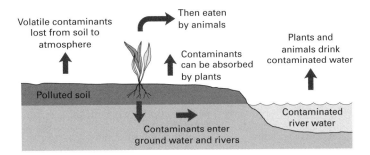

Fig. 7.1 Contaminants in the soil can find their way to other areas of the environment. This is why it is important to protect the quality of the whole environment, including the soil.

effect on organisms. This definition means that even relatively harmless contaminants, such as detergents, can pose a pollution hazard if present at high enough concentrations. Although many countries have recognized the need to protect air and water quality, it is only recently that soils have been afforded similar levels of protection, despite the fact that preventing soil contamination is a key factor in the provision of clean food and water. Figure 7.1 shows how soil contaminants can find their way into other areas of the environment.

How do soils become contaminated?

Contaminants can be released to the environment either from 'point sources', i.e. sources located in one spot such as landfill sites, or from 'diffuse sources' such as motor vehicle emissions. Listed below are some common sources of soil contamination, and Table 7.1 shows which contaminants are characteristic of certain industries:

♦ landfill sites (containing a wide range of contaminants, including organic wastes and metals);
♦ mine workings (metals);
♦ sewage sludge (which often contains high concentrations of metal);
♦ agricultural fertilizers (phosphorus and nitrogen can contaminate water sources);
♦ industrial wastes (metals, organic solvents, oils and chlorinated compounds);
♦ refineries and filling stations (hydrocarbons);
♦ pathogens (sewage sludge, meat rendering plants);
♦ radiation (nuclear power station accidents, military waste dumps, weapon tests and naturally occurring radon gas).

Since the earliest times, societies have used soil as a quick and convenient disposal route for rubbish. In many cases this has been the ideal way in which to recycle waste as it sometimes provides valuable plant nutrients.

Table 7.1 Pollutants characteristic of certain industries.

Industry	Contaminant
Metal mines, iron and steel works	Toxic metals: cadmium, lead, arsenic, mercury
Electroplating works and shipyards	Metals: copper, zinc, nickel
Gasworks and power stations	Combustible substances: coal and coke
Landfill sites and dock basins	Flammable gases: methane
Blast furnaces	'Aggressive' substances: sulphates, chlorides and acids
Chemical works and refineries	Organic pollutants: oil and tar residues, chemicals including chlorinated compounds
Industrial buildings	Asbestos

Source: Department of the Environment (DOE), Central Directorate on Environmental Protection (CDEP), Inter-Department Committee on the Redevelopment of Contaminated Land and Environmental Pollution Toxic Substances Division (EPTS) (1987) *Guidance on the Assessment and Redevelopment of Contaminated Land*, ICRCL 59/83. DOE, London.

However, urbanization and industrialization have increased the rate at which waste is generated, and altered the type of waste produced. In order to understand how contaminants behave in soil we can divide them into two groups: inorganic and organic compounds. We will start by looking at the main characteristics of some inorganic contaminants.

What are the major groups of inorganic contaminants?

For the purpose of this chapter, when we say that a chemical is inorganic it means that it is free of carbon–hydrogen bonds. Generally, as inorganic contaminants do not offer a very good food source to soil micro-organisms, they tend to persist in soil. This is because heterotrophic nutrition mainly depends on breaking the bonds that occur between the atoms in organic molecules. Some of the general properties of inorganic chemicals have already been covered in Chapter 3; here we will look at how three inorganic contaminants – metals, acid rain and radiation – behave in the soil.

Metals

When soil scientists talk of metal pollution they are usually referring to 'heavy metals', or those metals with densities greater than $5–6$ g/cm^3. The most common soil pollutants are: arsenic, cadmium, chromium, copper, lead, mercury, nickel and zinc. Most originate from industrial processes.

Table 7.2 Metal grouping based on toxicity to plants and animals.

Extremely toxic	Moderately toxic	Least toxic
Cadmium	Lead	Boron
Arsenic	Nickel	Copper
Chromium		Zinc
Mercury		

At very low concentrations, some of these metals are important micro-nutrients; in polluted soils, however, they are present at concentrations that are toxic to organisms.

Traditionally, one of the most important sources of metal contamination has come from the application of sewage sludge. Although sewage sludge has the potential to be valuable fertilizer (it typically contains 5% nitrogen and 4% phosphorus) it can also contain high metal concentrations. In the UK the association of sewage sludge with metal contamination dates back to 1937, when industry was given the right to dispose of some types of waste using the domestic sewage system. Today, many industrial processes are much cleaner. This has led to a gradual decline in the metal concentration of sewage sludge. Other sources of metal contamination include atmospheric deposition and agrochemicals. For example, the use of copper preparations to control fungal pests on vines has led to serious metal contamination in some areas of Europe.

Metals are toxic because they affect the permeability of cell membranes and disrupt the energy-producing functions of the cell. In the soil, high concentrations of metals can replace essential nutrients held on soil exchange sites. Potentially the most hazardous situations occur when a plant has the ability to tolerate high soil metal loads, but also forms a staple foodstuff. Under these circumstances metal contamination can be passed into the human food chain. In Table 7.2 metals have been grouped together on the basis of their toxicity to plants and animals.

Metal contamination can persist in the soil for very long periods. Basically, the reactions of metals are very similar to those of other cations (such as K^+) that were described in Chapter 3. The reactions of metal pollutants can be summarized as:
♦ adsorbed by some clay minerals;
♦ adsorbed by SOM;
♦ formed into metal crystals.
Although the metal load of sewage sludge in the UK is now in decline, sludge applications still need to be carefully monitored to ensure that the metal concentration of the soil remains below toxic concentrations (this will be discussed in more detail later).

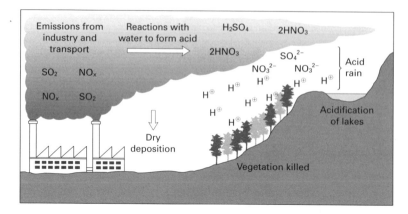

Fig. 7.2 Acid rain is formed when emissions rich in sulphur and nitrogen combine with water in the atmosphere to produce acid. The main sources of these polluting gases are motor vehicles and industry. (From Department of the Environment, Central Directorate on Environmental Protection and Royal Commission on Environmental Pollution (1985) *Controlling Pollution: Principles and Prospects.* HMSO, London.)

Acid rain

Acid rain is produced when gas emissions, rich in sulphur and nitrogen, dissolve in atmospheric moisture to produce acid. The main source of these gases is from the combustion of fossil fuels. The pH of rainfall in unpolluted areas is usually around 5.6; however, in some industrial areas it can fall as low as 2. Figure 7.2 summarizes acid rain formation.

Acidity has a major effect on many soil processes, including the behaviour of other contaminants such as metals, which tend to be more mobile in acidic soils. The relationship between soil acidity and metal pollution has been demonstrated in long-term experiments at Rothamsted Experimental Station (UK). Figure 7.3 shows the results from an experimental trial where the pH in several fields has been allowed to decline over a period of 140 years. The hay harvested from plots with the lowest pH was found to have 40 times the concentration of Al^{3+} as hay from limed plots where the pH had been maintained at 6–7. This concentration of Al^{3+} was 7–8 times the maximum tolerated by cattle. If the crop had been a foodstuff then the toxic concentrations of aluminium would have passed into the human food chain. For more information on soil acidity refer to Chapters 3 and 6.

Radiation

Radiation in itself is not necessarily a contaminant. Every day we are exposed to radiation, albeit at naturally occurring low levels. In an average lifetime 79% of our exposure will come from natural sources, 19% from medical procedures and 2% from power station accidents and bomb tests. Soils can become contaminated with radiation (i.e., have radiation above

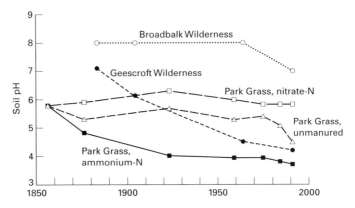

Fig. 7.3 The decline in soil pH in some non-limed, long-term field trials. The experiment illustrates the effect of some fertilizer applications, especially where nitrogen has been applied as ammonium, in promoting soil acidification. Soil pH values shown (measured in a 1 : 2.5, soil : water suspension) are for the surface (0–23 cm) soils from some of the classical experiments at Rothamsted Experimental Station. (From K. Goulding and B. Annis (1998) *Lime, Liming and the Management of Soil Acidity*. The Fertiliser Society, York.)

background levels) from a variety of sources, including weapon tests, poor waste disposal of contaminated material and accidents at nuclear power stations. High levels of radiation are toxic because it can deform DNA, the chemical code that controls cell replication. Once damaged, cell replication can become uncontrolled and can lead to the formation of tumours. The toxicity of radiation is related to:

♦ the extent to which it is retained in the body;
♦ its energy level;
♦ its half-life (or the time taken for half of the compound to decay);
♦ the part of the body in which it resides (some tissues are more suscept-ible than others to radiation damage).

Radioactive compounds behave like any other inorganic chemical, the only difference being that they are chemically unstable, and as a by-product of their instability they emit radiation. Three types of radiation can be detected: these are called alpha (α), beta (β) and gamma (γ) radiation. This distinction is important because it determines how damaging the radiation is and how easy it is to stop. For example, alpha radiation has the greatest potential to cause damage but is also very easy to stop; gamma radiation, on the other hand, is less damaging but is harder to contain. Figure 7.4 shows the main characteristics of α, β and γ radiation.

A major natural source of radiation is the emission of radon gas from rocks. Under natural conditions the gas simply disperses in the atmosphere to such low concentrations that it is no longer toxic. Problems occur when the gas seeps into the basements of poorly ventilated houses (see Fig. 7.5). In this situation it can build up to toxic concentrations. In theory the alpha radiation emitted from radon gas is relatively easy to stop by simple

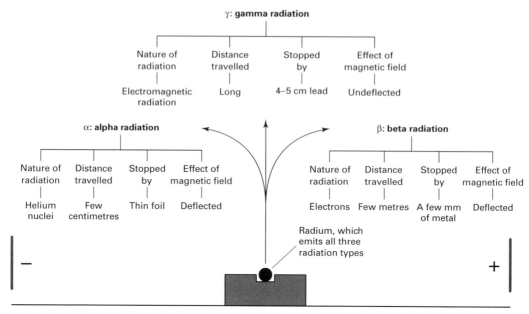

(NB: Distance travelled = in air)

Fig. 7.4 When a substance is 'radioactive' it emits particles that have the potential to cause damage. The main properties of each of the three types of radiation are shown. Radioactive substances can emit one or a combination of radiation types.

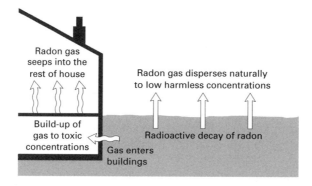

Fig. 7.5 Radon gas is a naturally occurring substance; however, problems occur when the gas builds up to toxic concentrations in poorly ventilated houses.

metal shielding. However, in most everyday situations the radioactive gas is inhaled directly into the sensitive tissues of the lungs, and long-term exposure can lead to lung cancer.

Other sources of radiation include fall-out from weapon tests and power station accidents. The radiation plume emitted from Chernobyl, for instance, contained isotopes such as ^{131}I, ^{134}Cs and ^{137}Cs. It was transported across Europe before being deposited as rain. It is estimated that 5000 people across Europe will die prematurely from the radiation emitted from Chernobyl.

The world's shortest course in organic chemistry . . .

Before looking at some organic contaminants, we need to understand a little about their chemistry. Organic molecules are constructed mainly of carbon and hydrogen atoms bonded together. Some organic compounds consist solely of carbon and hydrogen, while others have varying proportions of other elements present within their carbon framework. Organic molecules can range in size from small compounds, such as methane (which consists of one carbon atom and four hydrogen atoms), to very long and complicated structures that consist of thousands of interlinked atoms.

Before looking at some of the organic contaminants found in soil we first need to understand some of the basic properties of organic molecules. Our task is made easier in some respects because despite the wide range of organic compounds, their behaviour in soil is dominated by two characteristics, which are:
♦ molecular weight;
♦ polarity.

Molecular weight

Molecular weight is sometimes referred to as 'relative molar mass' and in very simple terms is the weight of the compound. Figure 7.6 shows how we can calculate the relative molar mass of two molecules by referring to weights listed in the periodic table.

Molecules with different molecular weights will have different physical characteristics. For example, simple organic compounds such as methane tend to be gases at room temperature, but as the molecular weight of a compound increases its tendency to form a gas decreases. Molecular weight has important consequences for soil pollution and remediation. Table 7.3 shows how increasing molecular weight affects the physical properties of hydrocarbons.

In terms of soil contamination, molecular weight is important because it will affect the ease with which organic contaminants can be removed from the soil. For example, given the right conditions, lighter compounds can often be removed from the soil by volatilization. Heavy compounds, on the other hand, can persist in the soil for many years.

Polarity

The other key characteristic is polarity. We have already seen the way inorganic compounds can become positively and negatively charged to form cations and anions (see Chapter 3). Some organic compounds can also develop ionic charges, but more commonly organic molecules experience subtle charge imbalances within their molecular framework. This means

Step 1: Obtain the relative atomic mass of each element in the compound

C	12
H	1
O	16
N	14

Step 2: Add up the relative atomic mass of each element in the compound–this gives
the molecular weight of the compound, e.g.:

(a) Water

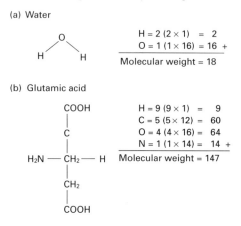

$$H = 2\ (2 \times 1)\ =\ 2$$
$$O = 1\ (1 \times 16) = 16\ +$$
$$\overline{\text{Molecular weight} = 18}$$

(b) Glutamic acid

$$H = 9\ (9 \times 1)\ =\ \ 9$$
$$C = 5\ (5 \times 12) =\ 60$$
$$O = 4\ (4 \times 16) =\ 64$$
$$N = 1\ (1 \times 14) =\ \ 14\ +$$
$$\overline{\text{Molecular weight} = 147}$$

Fig. 7.6 Chemists express the weight of a substance by calculating its relative molar mass.

Table 7.3 The relationship between carbon chain length and physical state.

Hydrocarbon	Number of carbon atoms	Physical state (standard temperature and pressure)
Methane	1	Gas
Ethane	2	Gas
Propane	3	Gas
Butane	4	Gas
Pentane	5	Liquid
Hexane	6	Liquid
Heptane	7	Liquid
Octane	8	Liquid

that although the overall molecule remains neutral, there are localized
areas of positive and negative charge. This property is referred to as
'polarity'.

If the molecule is simply constructed of carbon and hydrogen atoms it
will have very little polarity, but many organic compounds contain other
elements such as oxygen, chlorine, phosphorus and sulphur, which are
interspersed throughout the main body of the molecule. Different atoms

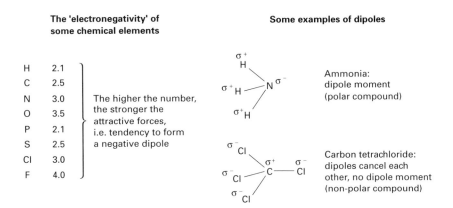

Fig. 7.7 The formation of dipoles and dipole moments is shown using two simple chemicals: ammonia and carbon tetrachloride.

have different electrical characteristics. If two atoms with similar charges, such as two carbon atoms, are joined together there is no separation of charge and the molecule is said to be neutral. However, if two different atoms with different electrical properties are joined together they produce a molecule that has a positive end and a negative end. These areas of charge separation are called 'permanent dipoles'. Sometimes dipoles cancel each other out, so that the molecule has no overall charge imbalance; in other cases the molecule will have localized areas of positive and negative charge; these are referred to as a 'dipole moment'.

Dipoles are usually shown using the symbol δ with the corresponding + or − sign. The relatively small charges associated with dipoles are not as strong as the ionic charges associated with cations and anions but they are sufficient to affect how the molecule behaves in soil. Figure 7.7 shows two molecules: one is non-polar, the other polar.

The simplest way to understand why the polarity of a compound is important is to say that compounds with similar polarities tend to stick together and those with opposite polarities repel. The best way of illustrating this is to look at Fig. 7.8. If you place a scoop of margarine in a bottle of water and shake, the margarine remains unchanged. Now if you add olive oil to the bottle, shake and leave to stand, you find that the margarine has mixed with the olive oil, which has acted as a solvent. This is because margarine and olive oil have similar polar characteristics. If the bottle is left to stand overnight, so that the two solvents, the water and olive oil, separate, you will find the margarine has dissolved in the olive oil to form a layer of liquid that overlies a layer of water. The margarine and olive oil are referred to as the 'non-polar phase' and the water the 'polar phase'. Chemists can use this process to separate liquids that have different polarities.

Soils naturally contain all kinds of chemicals, some polar and some non-polar, just like the margarine in our example. When a contaminant is

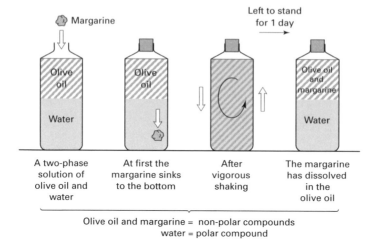

Olive oil and margarine = non-polar compounds
water = polar compound

Fig. 7.8 We can demonstrate the way contaminants bond with soil chemicals that have similar polarities by mixing some household substances together in a simple experiment.

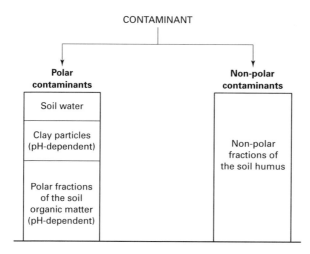

Fig. 7.9 If we divide contaminants on the basis of their polarity we can predict how they will be partitioned between the mineral, inorganic and organic fractions of the soil.

deposited in the soil it will bind to chemicals that have similar polar characteristics. It is for this reason that many non-polar organic contaminants are often associated with the non-polar fractions of the soil organic matter pool. In Fig. 7.9, the main soil fractions have been separated on the basis of their polarity.

Molecular polarity can also affect how easily a chemical is degraded. Very often micro-organisms attack large molecules where there is a dipole. It is for this reason that chemicals such as heavy hydrocarbons, which have a combination of high molecular weights and non-polar characteristics, can be amongst the most persistent organic contaminants.

What are the main types of organic contaminants in soils?

Having considered the general properties of organic molecules, we now need to look at how these characteristics affect the behaviour of organic contaminants in the soil. We will consider three organic contaminants: pesticides, petroleum products and chlorinated hydrocarbons.

Pesticides

Modern agriculture relies on the control of insects, fungi and plant pests by the applications of biocides. This is not a new development – before the development of modern synthetic pesticides pest control was achieved using general poisons such as arsenic, metal salts and nicotine fumigants. The development of synthetic organic pesticides started in 1939 with the synthesis of DDT, followed shortly by the production of the herbicide 2,4-D in 1941.

Pesticides can be divided into a number of chemical groups, but they are evaluated on the basis of three main properties:
♦ their ability to control the pest;
♦ their selectivity (the best pesticides attack only the target organism);
♦ their persistence in the environment.

DDT is probably the most famous (or infamous) insecticide. It belongs to a family of compounds called the organochlorines (OCs). If measured purely on the basis of short-term human welfare, it can be regarded as one of the most beneficial chemicals ever synthesized. This assertion stems from the use of DDT in the 1950s to control mosquitoes, which are the vectors of several major tropical diseases such as malaria and yellow fever. However, many OC insecticides, such as DDT, suffer from the fact that they are not selective (they reduce the numbers of beneficial insects along with the pest species) and that they are persistent. It is this tendency to persist in the environment and accumulate in the fatty tissues of animals higher up the food chain (a process called 'biomagnification') that finally led to restrictions on their use. Figure 7.10 shows how biomagnification can lead to pesticide accumulations in higher animals.

Today, the most popular group of agricultural insecticides are the organophosphates. Compared with OCs, they have low environmental persistence times; unfortunately, though, they also tend to be much more toxic to mammals. Some farm workers who have experienced long-term exposure have suffered neurological damage. In some ways this is not surprising, as the initial driving force behind the development of this group of chemicals was not to control agricultural pests, but to produce nerve gas for military purposes. Other insecticide groups include the carbamates and pyrethroids, both of which are modelled on naturally occurring neuropoisons found in plants.

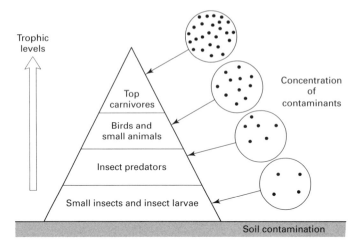

Fig. 7.10 The concentration of non-polar contaminants in the fatty tissues of animals increases as you progress up the food chain. This process is referred to as biomagnification.

Unlike insecticides, which remove an unwanted species feeding directly upon our food reserves, herbicides are designed to reduce competition by removing unwanted plants (weeds). The most popular herbicides belong to the phenoxyacetic acid group. Generally, herbicides do not persist in the environment and most have low mammalian toxicities. Some, however, such as the defoliants 2,4,5-T and 2,4-D (better known as 'Agent Orange' after the orange-coloured band on the drums it was stored in) were used extensively during the Vietnam war and have been implicated in ill health and birth defects.

Figure 7.11 shows the persistence of some common pesticides.

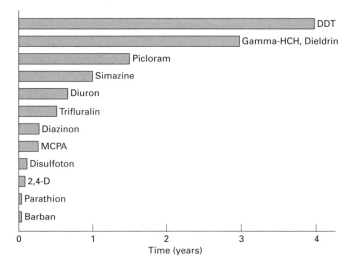

Fig. 7.11 The persistence of some common pesticides in soil. (From I. J. Graham-Bryce (1981) The behaviour of pesticides in soil. In: *The Chemistry of Soil Processes*, eds D. J. Greenland and M. H. B. Hayes. John Wiley, Chichester. © John Wiley & Sons Ltd. Reproduced with permission.)

Table 7.4 Crude oil products with their respective carbon chain length and physical state.

Product	Length of carbon chain	Physical state (STP)
Refinery gas	1–4	Gas
Light petroleum	5–6	Liquid
Light naphtha	6–7	Liquid
Petrol (gasoline)	5–12	Liquid
Paraffin (kerosene)	12–18	Liquid
Gas oil	18–25	Liquid
Lubricating oil	20–34	Liquid
Paraffin wax	25–40	Solid
Bitumen	>30	Solid

Source: E. N. Ramsden (1994) *'A' Level Chemistry*, 3rd edn. Stanley Thornes, Cheltenham.

Petroleum-based products

Table 7.4 shows how crude oil can be divided into several petroleum products.

Many petroleum-based compounds are included under the term 'polyaromatic hydrocarbons' or PAHs. The USEPA (United States Environmental Protection Agency) and the EC list 16–20 petroleum-based hydrocarbons that they consider 'priority pollutants'. Major sources to the environment include the incomplete combustion of fossil fuels and leaky petrol storage tanks. They are predominantly non-polar compounds with low water solubilities that tend to bond with soil organic matter. The lighter compounds tend to be quite volatile, and can be removed from soil using various extraction procedures. Heavier compounds, however, pose a more serious threat because they do not volatilize easily, so remain in the soil.

Chlorinated compounds

Chlorinated compounds pose a particular risk in the environment as they tend to be toxic and have long persistence times. Most concern has focused on a range of chlorinated chemicals called PCBs (polychlorinated biphenyls) which were first developed in 1929. Since then 1–2 million tonnes of PCBs have been produced. They are used mainly as industrial fluids in capacitors, transformers and in the plastics industry. There are now over 209 different PCB compounds. Problems with environmental persistence were first detected in 1966 when researchers, looking for signs of DDT accumulation, found their results were being obscured by other compounds similar to DDT. These were later identified as PCBs. This led to concern about the environmental risks attached to these compounds, particularly the potential for biomagnification in the human food chain. PCBs are believed to cause skin irritation and liver damage. During the 1970s the UK banned the use of PCBs and although world production has ceased, it is estimated that 60% of the PCBs produced have yet to be disposed of.

What can happen to organic contaminants once in the soil?

Once deposited in the soil, organic contaminants can undergo the following reactions:

♦ Volatilization: compounds with low molecular weights, which are not strongly absorbed by soil particles, can be lost to the atmosphere. For example, many PCBs tend to be quite volatile chemicals, so once exposed at the soil surface they can often be removed by volatilization.

♦ Leaching: if the compound is water soluble it may be removed from the soil by leaching. This can represent a serious pollution hazard if the ground water is used as a drinking-water source.

♦ Adsorption: contaminants can become adsorbed by various soil constituents, such as clay minerals and soil organic matter, depending on their polarity.

♦ Biological absorption: some contaminants are absorbed and held intact or broken down by plants and animals.

♦ Microbial degradation: this is the most important pathway by which the potentially harmful effects of most contaminants are mitigated as the compound is mineralized to non-toxic substances.

♦ Chemical degradation: some chemicals are broken down and rendered harmless by chemical processes such as hydrolysis, oxidation and exposure to light.

How can we assess whether or not a soil is polluted?

In order to assess the risk that a contaminant poses we have to carry out a 'risk assessment'. It is only when we have done this that we can we say if a contaminant has become a pollutant. Let us return to the definition of 'pollution' we made at the start of this chapter. We said: 'Contamination becomes pollution once it has an adverse effect on organisms'. In order to determine this, 'target organisms' (which can be either plants or animals) are exposed to increasing concentrations of the contaminant. The effects of the contaminant on the organism are then noted. This type of experiment is called an 'ecotoxicology test'. The results can then be expressed as a dose–response curve, like the one shown in Fig. 7.12. Although this is an example of a plant response to increasing metal concentrations, it could have been any contaminant and any test organism; for example, we could have used changes in microbial respiration as a measure of toxicity.

Results from ecotoxicology tests are often open to alternative explanations, depending on the conflicting motives of environmentalists, industrialists,

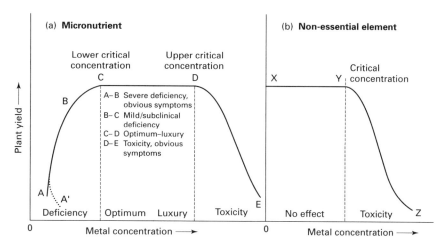

Fig. 7.12 The effects of a contaminant can be shown by exposing a target organism to increasing concentrations of the contaminant; the results can be expressed using a dose–response curve. (From B. J. Alloway (1990) *Heavy Metals in Soils*. Blackie, Glasgow.)

consumers and governments. For example, we could quite easily vary the acceptable permitted metal concentrations by using an organism that is more or less sensitive to metals; this has made setting pollution limits a difficult issue.

The UK's pollution limits have been set by the ICRCL (Inter-Department Committee on the Redevelopment of Contaminated Land) and provide 'critical concentrations' for certain industrial contaminants, above which the soil is classified as contaminated. Two sets of values are listed, first a 'trigger value' that alerts us that the soil is contaminated, and secondly an 'action threshold' above which the soil is regarded as polluted and requiring some form of remediation. Table 7.5 shows some ICRCL trigger and action threshold concentrations for a range of pollutants.

Table 7.5 ICRCL trigger and action threshold concentrations for a range of pollutants.

		Threshold values (mg/kg)	
Compound	Proposed land use	Trigger	Action
PAHs	Domestic gardens	50	500
Phenols	Domestic gardens	5	200
Free cyanide	Domestic gardens	25	500
Acidity	Domestic gardens	pH 5	pH 3

Source: Department of the Environment (DOE), Central Directorate on Environmental Protection (CDEP), Inter-Department Committee on the Redevelopment of Contaminated Land and Environmental Pollution Toxic Substances Division (EPTS) (1987) *Guidance on the Assessment and Redevelopment of Contaminated Land*, ICRCL 59/83. DOE, London.

Table 7.6 Pollution pathways.

◆ Soil	→	crop	→	human			
◆ Soil	→	crop	→	livestock	→	human	
◆ Soil	→	livestock	→	human			
◆ Soil	→	surface water	→	human			
◆ Soil	→	surface water	→	fish	→	human	
◆ Soil	→	ground water	→	human			
◆ Soil	→	air	→	human			

Although reference data can help us to answer the question, 'Is the site polluted?', they do not represent a detailed risk assessment. Further work is often needed to establish the 'bioavailability' of the pollutant and the way in which it could enter the human food chain. There are usually a number of routes by which the pollution can cause damage to organisms. Several 'pollution pathways' are shown in Table 7.6.

The problems of risk assessment are best illustrated if we use an example. Imagine an old industrial site that has recently been purchased by a building contractor. The contractor hopes to redevelop the land for housing. Before this can be done he needs to assess whether the site is polluted. The contractor has now asked you to carry out a pollution risk assessment of the area. How do you think you would go about doing this?

Any risk assessment will consist of several steps. We will now go through them in order, each posed as a series of questions:

Step one: What is the history of the site?   The description of the site in the brief mentioned 'industrial', but this could mean anything. Before we start the survey we need to investigate the history of the site so that we can build up some idea of which contaminants might be present. For instance, we need to know what types of industry have used the site in the past, for how long and where they were located. We also need to know something about the soils at the site, especially their texture and hydrological properties. We can use this information to focus on the most probable pollution pathways.

Step two: What is the concentration and distribution of contaminants?   By now we should have a better understanding about the problem we are facing. It would be impractical to analyse for every possible contaminant so a selection is made depending on the past history of the site, the proposed redevelopment use and the potential toxicity of the contaminants. In order to survey the site we need to have a clear idea about how many samples need to be taken, which areas to take them from and how they should be stored. (For more information on sampling techniques see Chapter 5.) Once the soil samples have been taken they can be sent to a laboratory and analysed.

Step three: Is the soil contaminated or polluted?   This is the crucial question, and the one the property developers will be anxious to have an answer to. Once we have the results back from the laboratory, we can consult reference tables that list the maximum permissible concentrations for the range of contaminants we have investigated.

How can we remediate a polluted site?

Let us return to our hypothetical property developer. We have presented the bad news – the site is polluted and the pollution poses an unacceptable health risk. What are the options?

There are two things the developer could do. First, and by far the quickest and cheapest, is to use the land for another purpose. By altering the land use we can often change the range of pollution pathways. For example, rather than building houses we could build a car park instead, where the soil surface is effectively sealed. This process is referred to as 'soil capping'. The other, more expensive option is to remediate the land. Let us assume the property is in a desirable location and its potential value is high, so that it makes commercial sense to clean it up. How can we do this?

The most important factors to consider are: what types of pollutants may be present, and at what concentrations. There are a number of processes that can be used to remediate the soil. These include: microbial action, immobilization within biomass and soil venting.

Microbial action

Easily degradable organic compounds can be removed from the soil by creating the optimum chemical and physical conditions for microbial activity. This can be achieved by increasing aeration by ploughing and adding fertilizer. This method is sometimes referred to as 'land farming' and is only suitable for organic compounds that can be mineralized. Simply leaving soil processes to take their natural course is the single most important route by which organic contaminants are lost from the soil.

One example of how land farming can be used is to remediate soils polluted with petroleum products. Many of the chemicals in crude oil are biodegradable, given the right conditions. In order to remediate soils contaminated with hydrocarbons we have to ensure that the soil conditions are optimized for high rates of biological activity. In the case of large spills of light petroleum products, much of the pollution can be removed directly by digging several shallow drainage holes. Once the petroleum has moved through the soil matrix into the drainage holes it can be pumped out. Heavier compounds tend not to penetrate the soil to the same extent. Here, the best approach is to let the petroleum solidify slightly, then remove

A shoddy gamble that never stood a chance

The whole site consists of some 250 acres, of which about 50 have been taken up by the Dome itself. During the time that the north bank was being developed, the Greenwich peninsula site was also available for redevelopment and was, in fact, considered by a large number of potential developers. In the past week I have been telephoned by two friends who happened to have been involved separately in looking at the site. One was most impressed when he flew over it in the late 1980s, the other discussed a possible scientific use for it in the early 1990s.

Both proposals fell down because the site was too heavily polluted by the residues of more than 100 years of gas production, of chemicals, of oil refining, of steel production. On the second occasion, the developer had even spent his own money to put down some boreholes, which established the existence of poisonous metals, including lead and arsenic, going down to a level of 30ft. There are also toxic and carcinogenic chemicals.

This developer estimated that the greater part of the poisoned soil would have to be removed. In all it could amount to a quarter of a billion cubic feet. There would be great difficulty, as well as cost, in finding a place in which to dump it. Removing the land which supports the rusting seawall, which already looks fragile in places, would probably mean that the wall would have to be replaced. One estimate was that the building of a new wall, on its own, would equal in cost the whole economic value of the site.

Any plan for redevelopment has to face this difficulty; if the developers of a decade ago found it prohibitive, what has changed? The construction even of the Dome itself did not include excavation of the contaminated soil.

The Dome site was scraped to a depth of 18in, a membrane was laid, and covered in hardcore and concrete. Greenwich council would give the Dome only a limited, 12-month planning consent, for exhibition purposes. The rest of the site has not even been scraped and covered as far as one can see when one walks around it; scrape and cover is not safe enough for residential use.

Pollution is now recognised as a much more serious environmental and health hazard than was appreciated even in the early 1990s. The contamination of the Greenwich site will already have spread through the water table to adjoining areas. There is a responsibility on the owner in law to compensate people injured by this contamination, or those whose properties are reduced in value. It is doubtful whether the whole site, or the Dome site on its own, could be sold to anyone without a guarantee against the damages which might arise from the underlying pollution.

Four questions had to be answered; they reflect on both the present Government and on its predecessor. How did they think that the pollution problem, which was known to everyone who had ever dealt with the site, would be solved? Why did they proceed with the extremely expensive plans for the Jubilee Line and the Dome without having established, that the site could, in fact, be cleared? Why did the Government proceed without having secured a major developer, who had been sought unsuccessfully for a decade? Why did it go ahead without having a development plan in place? Even now, the Government will not know who will buy the site or what will be done with it. It is no defence to argue that the Jubilee Line has at last been constructed, or that it is a very handsome architectural achievement.

The location of the North Greenwich station was determined by the site. Unless the problems of the site can be solved, the line itself may be in the wrong place.

Nomura withdrew from its intended purchase because it thought that it had not been given adequate information. What one must wonder is whether anyone, at any point, had a proper grip on these interlinked projects. From the point at which it became redundant for the purposes of the gas board, there was a heavily polluted, but geographically important, site available in North Greenwich. The potential cost of clearing the pollution was greater than its development value. Various projects were considered. They all failed to come to anything.

There was another project to provide a new Underground link between East London and Westminster. Somehow that project became entangled with the Greenwich site, without the pollution problem having been resolved. On top of that came the National Lottery, with its vast funds outside Treasury control, in essence irresponsible money. These funds were made available to the Millennium Commission, which decided to spend them on the Dome. This crazy bamboo scaffolding was ill-designed to bear any weight and it has fallen down.

There are wider issues. The Millennium Commission has spent money on keeping the Dome open which should have been spent on proper purposes. Lottery funds are public money, and need to be controlled as such.

Plainly there has been a breakdown in the effectiveness of regional planning. The specific administrative breakdown of the Dome has been very serious. But I also wish that politicians would apologise, and do so frankly. I would like a proper apology from Michael Heseltine, the original begetter of the catastrophe, from Peter Mandelson, its enthusiastic sponsor, and, above all, from Tony Blair.

As he says, it happened on his watch; why will he not say that he is sorry?

Comment@the-times.co.uk

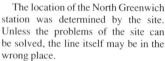

William Rees-Mogg

Fig. 7.13 This article taken from *The Times*, London (2 October 2000) neatly sums up some of the problems polluted soils cause to the redevelopment of old industrial sites. (© William Rees-Mogg/Times Newspapers Limited, London 2000.)

the contaminated surface layers of the soil. In both cases the remaining petroleum can be removed from the soil by microbial activity. We can optimize this by adjusting the pH to 6–7 with lime, adding fertilizer and by ploughing. It may take several years before pollution concentrations are reduced below action thresholds. Soils contaminated with light oils can usually be remediated within 1–2 years, while sites contaminated with heavier compounds may take many years to clean. In some cases, where the pollution is very severe, the only practical way of remediating the site will be to remove the contaminated soil completely. The news article in Fig. 7.13 describes some of the problems facing the developers of the Millennium Dome site in London.

Immobilization within biomass

Some trees and plants are tolerant of some forms of pollution. Sometimes it is possible to grow plants on the polluted site. For example, metals can be taken up by the trees and immobilized within their tissues. Once mature, the trees can then be felled and the wood utilized in ways in which metal contamination does not pose a health risk, for example construction. In other cases specialized plants, which have the capacity to survive in soils that have high metal concentrations, can be grown on the site. These plant species have been called 'hyper-accumulators' and under natural conditions are often found in soils that have formed from metal ores. Hyper-accumulators work by immobilizing metals within their tissues. Repeated cropping, followed by harvest and incineration, gradually reduces the metal concentration of the soil by concentrating it in a relatively small volume of plant tissue, as shown in Fig. 7.14.

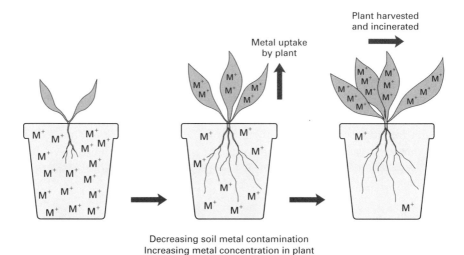

Fig. 7.14 The use of hyper-accumulator plant species to concentrate metal pollution.

Soil venting

A stream of air or steam can be passed under pressure through the soil. This method can be used to remove volatile toxins such as light hydro-carbons, for example gasoline.

What can we do if remediation is not possible?

In some cases it may not be possible to remediate the site. The contaminant may be too toxic or the concentrations so high that all of the methods described above are ineffective. In the case of serious pollution the main aim is to protect human health by removing the risk as quickly as possible. There are several ways this can be achieved.

Incineration is a way of removing organic pollutants. The process is rapid and is capable of treating large volumes of soil. The contaminated soil is simply removed and incinerated. Problems include high costs and emissions. Otherwise, the contaminated soil can be combined with a binding agent such as concrete; it then can be broken up and dumped or used for road foundations. This method can be used for organic and inorganic contaminants. In the case of very toxic substances the soil can be set in glass using a process called 'vitrification'. This process is expensive and limited to those contaminants that pose serious health risks such as high-level radiation. Although these methods are effective at preventing the contamination spreading to the wider environment, they all involve the loss of a valuable non-renewable resource – the soil.

How can we prevent soil pollution?

In many cases with careful management we can prevent contaminants becoming pollutants. Generally, it is far cheaper to prevent pollution in the first place than clean it up afterwards. Society has always used soil to dispose of its waste. It is a lasting testament to the diversity of the soil microbial population and their ability to break down contaminants that very few diseases can be contracted directly from the soil. Like many societies we have become accustomed to burying our waste and then promptly forgetting about it. However, there is a growing recognition that the soil is a non-renewable resource that needs to be carefully managed. We will look at how we can manage landfill sites and sewage sludge applications.

Landfill

Compared with other EC countries, the UK disposes of a disproportionate amount of its waste as landfill and still practises co-disposal, where by certain industrial and domestic wastes are buried together. Society should

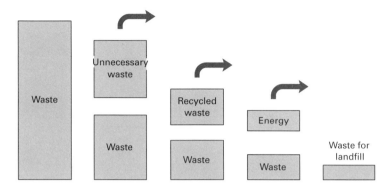

Fig. 7.15 The most effective way of managing waste is to take steps to prevent creating it in the first place. This reduces both waste and the amount of land given over to landfill.

regard landfill as the last option. Before using the soil as a dumping ground we should ask whether the waste could have been prevented in the first place (for example, by reducing excess packaging, recycling schemes or using the waste to generate energy). It is only when we have exhausted these three possibilities that landfill should be considered as a waste disposal option. These options are shown in Fig. 7.15.

Landfill sites can suffer from several problems if not managed properly. For example, as organic material is broken down, slumping and gas emissions, such as methane, nitrous oxide and carbon dioxide, can occur. In addition, if the site is not sealed properly, leachates that contain inorganic and organic contaminants may leak to ground-water sources. Figure 7.16 shows some of the reactions that can occur during the decomposition of organic material in landfills and the resulting contamination plume.

Many of the problems associated with pollution from landfill sites can be remedied by simply regulating the type of waste that is buried. In many EU countries landfill waste is limited to inert material that will not react further once buried (for example, glass and building rubble). Once the site has been filled, its surface is sealed with an impermeable cap so that the leaching risk is minimized. Because the site does not contain organic material there is little risk of gas build-up or toxic leachates moving to ground water. This method of managing landfill has been termed the 'dry tomb' approach.

For organic waste a 'bioreactor' landfill site can be used. Here, organic material is placed in a large pit that has been lined with either plastic or clay, which prevents leachates moving down into ground-water sources. Leachates that collect at the bottom of the sealed pit are pumped back to the surface and then recirculated. Gases such as methane can be removed from the site using vents. This reduces the explosion risk and can sometimes be used as a source of energy. After a period of several years much of the mineralizable matter will have been lost. The site can then be capped

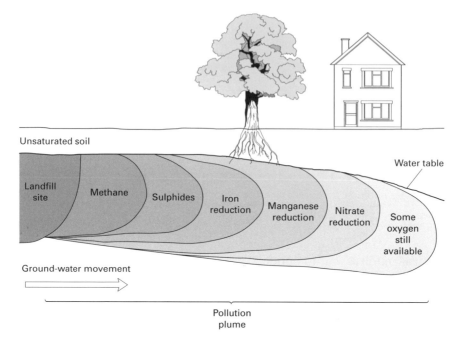

Fig. 7.16 The decomposition of organic waste in non-lined landfill sites can create a pollution plume that can contaminate the surrounding area. (From Department of the Environment, Central Directorate on Environmental Protection and Royal Commission on Environmental Pollution (1985) *Controlling Pollution: Principles and Prospects*. HMSO, London.)

and used for redevelopment or the material removed and used as compost, before new material is added to the site. Figure 7.17 compares a bioreactor site with a conventional landfill.

Sewage sludge

In order to manage sludge disposal in a way that minimizes the pollution risk we need to know:
◆ the metal load of the sewage sludge;
◆ the initial metal load and pH of the soil;
◆ the maximum soil metal concentrations that are still considered safe.
Soil, in many respects, is the ideal route to dispose of sewage sludge. In order to prevent the concentration of heavy metals reaching pollution concentrations several countries have published maximum permitted metal limits for soils receiving sewage sludge. These are shown in Table 7.7.

Unlike many of the other soil management issues we have looked at, pollution control is a political as well as a practical issue. Some countries have opted for a more cautious approach than others. For example, the Dutch have two sets of values, labelled A and C. The A values indicate when a soil is contaminated and the C values indicate when some kind

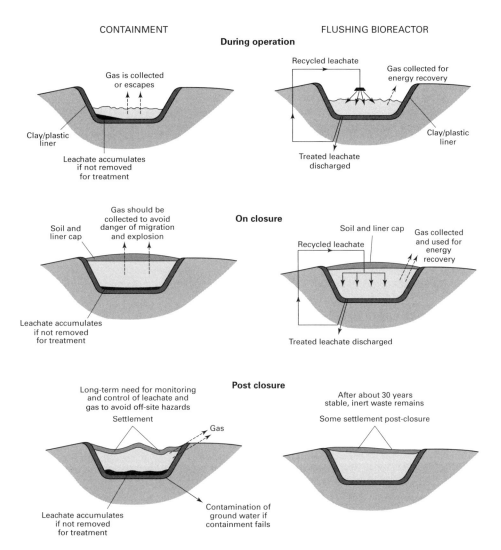

Fig. 7.17 Containment and flushing bioreactor landfill sites. (From Department of the Environment, Central Directorate on Environmental Protection and Royal Commission on Environmental Pollution (1985) *Controlling Pollution: Principles and Prospects*. HMSO, London.)

of remediation is necessary; however, many of the Dutch C values are *below* the acceptable metal concentrations of the USA.

Generally, the main aim of managing metal-contaminated soil is to stabilize the metal within the soil. We can do this by ensuring that the pH of the soil remains at or above 6–7 with regular lime applications because metals tend to be less mobile in alkaline environments, and by preventing waterlogging, as most metals are also less mobile when in an oxidized form. Finally, organic matter can be used to bind metals within the mineral fabric of the soil. There are always some exceptions to these general rules,

Table 7.7 Some typical threshold values (ppm) for metal concentration in agricultural soils (pH 6–7) in a range of countries.

Country	Cd	Cu	Cr	Pb	Zn	Hg
EU	1–3	50–140	100–150*	50–300	150–300	1–1.5
France	2	100	150	100	300	1
Germany	1.5	60	100	100	200	1
The Netherlands						
A	0.8	36	100	85	140	0.3
C	12	190	380	530	720	10.0
United Kingdom	3	135	400	300	300	1
United States	20	750	1500	150	1400	8

*Values now withdrawn.
Source: S. P. McGrath, A. C. Chang, A. L. Page and E. Witter (1994) 'Land application of sewage sludge: scientific perspectives of heavy metal loading limits in Europe and the United States', *Environmental Review* **2** (NRC Press, Ottawa).

and in the case of metal pollution the exception is arsenic, because it tends to be more mobile in alkaline soils. It is therefore important, if the site is believed to contain arsenic compounds, that soil pH is measured before lime is applied.

Essential points

◆ Soils are contaminated when they have elevated levels of chemicals, usually as a result of human activity, compared with soils that are regarded as being in pristine condition. Contamination is regarded as pollution once the contaminant has an adverse effect on human health or the biological activity of the soil.
◆ Contaminants can be divided into inorganic and organic compounds.
◆ Inorganic contaminants include metals, acid rain and radiation.
◆ Organic contaminants are composed primarily of carbon and hydrogen atoms, bonded together into molecules. Their molecular weight and polarity determine how they will behave in the soil.
◆ Organic contaminants include pesticides, polyaromatic hydrocarbons, and chlorinated compounds. The main route by which organic chemicals are lost from the soil is through microbial degradation. Other routes include volatilization, leaching, adsorption and absorption by plants and animals.
◆ We can assess the concentration at which a contaminant becomes a pollutant by conducting an ecotoxicology test.
◆ Sites can be remediated by encouraging the degradation of the chemical by creating the right biological conditions, immobilization within the biomass, soil venting or incineration.

♦ It is better to prevent soil pollution happening in the first place by controlling the application of waste to soil so that the concentration of contaminants is kept below pollution trigger values. One problem with this approach is that there is often little agreement between countries on what constitutes pollution as opposed to contamination.

2   What is soil erosion?

Soil erosion: the nature and scale of the problem

Soil erosion is a natural phenomenon. All natural landscapes are subjected to some degree of erosion. In many areas the movement and deposition of soil is an essential component in maintenance of soil fertility. One of the best examples of this situation was in Egypt, where soil eroded in the Ethiopian uplands was deposited on the Nile delta. This process maintained the fertility of the soil for generations (the construction of dams to control the flow of the Nile has now altered this balance).

Erosion problems occur when the rate of erosion is increased above natural levels, usually as a result of poor agricultural management. In the humid tropics where rainfall can be intense, the signs of soil erosion are obvious when rivers turn thick and muddy following a heavy downpour. Even in the temperate latitudes, although the outward signs that erosion is occurring may be less dramatic, soil can still be lost from farmland at economically important rates. Soil erosion not only results in the removal of topsoil (where all the nutrients are concentrated) but also turns the soil into a troublesome pollutant once it is discharged into watercourses.

In 1960 the US Soil Conservation Service set the maximum acceptable loss of soil in the USA at 5 t/yr per hectare. It is now estimated that up to 80% of the world's agricultural land is subject to some form of erosion. The problem is illustrated by the fact that soils on average form at the rate of 1 t/ha each year, yet estimated losses from agricultural lands are 30 t/ha in Africa, Asia and South America and 17 t/ha per year in Europe and the USA. The effects of erosion can be summarized as follows:
♦ removal of topsoil and nutrients;
♦ reduction of rooting depth and water storage;
♦ increased silt load of rivers, leading to flooding, blocked drains and the silting-up of reservoirs;
♦ eutrophication of watercourses by sediments that contain adsorbed nutrients;
♦ pollution of watercourses by sediments that contain adsorbed pollutants.

One of the best examples of wind erosion occurred in the USA during the 1930s, in the region of the Great Plains. After a succession of dry years, starting in 1934, the dry soil was picked up by the wind. The resulting thick dust cloud, which was composed of valuable topsoil from the region, was transported as far as New York City and the eastern seaboard. This disaster became known as the 'dust bowl'.

Soil loss: the mechanisms of erosion

In order to combat erosion we need to understand the mechanism, as once this is done we can begin to manage the problem. We can divide erosion into two processes:

♦ the detachment of soil particles;
♦ the transport of dislodged particles.

The detachment of soil particles

In order to break up soil aggregates energy is needed. This usually comes from wind or water. For example, the impact of raindrops, commonly called 'rainsplash', can be especially destructive in tropical thunderstorms. Rainfall or wind energy, and its potential to cause erosion, is sometimes referred to as its 'erosivity'. In the case of rainfall its erosivity is highly dependent on its intensity and duration.

Just because an area suffers from heavy rainfall it does not automatically mean soil erosion will be a problem. This is because some soils are more susceptible to erosion than others; in other words we must consider what is termed a soil's 'erodability'. A soil's ability to withstand rainsplash depends partially upon its textural characteristics. Soils with a high proportion of sand and silt are the most vulnerable to erosion. The erodability of soils can be reduced to some extent by additions of organic matter that helps bind soil particles into stable aggregates.

The transport of dislodged particles

The second stage of the erosion process is the transport of dislodged material away from its original site. The main controlling factor is surface 'run-off'. Run-off occurs when the rainfall intensity is greater than the infiltration capacity of the soil. In regions that suffer from high-energy rainfall, infiltration rates can be greatly reduced by processes such as soil crusting. Once overland flow has started to occur the water rarely runs in a uniform sheet across the landscape but instead is funnelled into small, naturally occurring depressions. Once this happens, flow becomes faster and more turbulent, so that more soil is scoured from the channel. As the water cuts downwards, it carves a small channel called a rill. These are not usually permanent features as they can be easily ploughed out. However, when coupled with the right hydrological conditions rills can deepen into 'gullies'. Gullies are deep channels that are permanent landscape features,

Fig. 7.18 Rill erosion in the Peak District. The loss of vegetation on the footpath has led to rill erosion developing on the steeper sections of the path.

Fig. 7.19 Rill erosion in Kenya. As a result of intense tropical thunderstorms soil erosion can develop into a serious problem unless measures are taken to prevent it.

which cannot be rectified by ploughing. Gullies represent a serious loss of soil from the area where they have formed. The photographs in Figs 7.18 and 7.19 show rill erosion on a footpath in the Peak District, England, and serious rill erosion in Kenya.

Table 7.8 MAFF erosion risk assessment scheme.

Soil texture	Steep slopes >7°	Moderate slopes 3°–7°	Gentle slopes 2°–3°	Level ground <2°
Sandy loam	Very high (high)*	High (moderate)*	Moderate (lower)*	Slight
Silty clay loam	High (moderate)*	Moderate	Lower	Slight
Other textures	Lower	Slight	Slight	Slight

*Where average rainfall is less than 800 mm, the risk class in parenthesis applies.
Source: MAFF (1999) *Controlling Soil Erosion: A Field Guide for an Erosion Risk Assessment for Farmers and Consultants*. MAFF Publications, London.

How can we assess the erosion risk?

We can determine how vulnerable a site is to erosion by considering its erosivity and erodability. We can assess the erosivity of the area by consulting weather data, particularly the number of intense storms during the year. Secondly, we can consider the erodability of the soil by considering its texture and organic matter content. This is often coupled with site characteristics, such as slope and whether the soil is exposed or has vegetative cover. In the UK, MAFF has used factors such as these to construct a risk assessment scheme for farmers (Table 7.8). The aim of this scheme is to enable farmers to identify areas of their farm that could potentially suffer from erosion.

How can we prevent soil erosion?

If the site is vulnerable to erosion we can modify our farm management to minimize the risk. MAFF has produced a set of guidelines for farmers (Table 7.9). These can be used to identify farming practices that are not suitable for areas suffering from varying degrees of erosion risk.

As well as matching the ideal crop for the prevailing soil conditions, farmers can also modify their land management to reduce the risk of erosion. Summarized, these rules are: try to reduce the time the soil surface is left exposed: break up slopes so that the energy of any overland flow is kept to a minimum, and maintain levels of soil organic matter (SOM) so that soil aggregates remain stable. These rules can be formulated into a set of general principles that have worldwide application:

♦ Do not create unnecessary fine seedbeds.
♦ Keep the amount of time land is left bare to a minimum.
♦ Reduce compaction so that water infiltrates quickly.

Table 7.9 Risk assessment in relation to land use.

Low risk	Moderate risk	High risk/Very high risk
No limitation	Oilseed rape Early-sown cereals Spring-sown linseed	Oilseed rape Early-sown cereals Spring-sown linseed Potatoes Sugar beet Vegetables Pigs

Source: MAFF (1999) *Controlling Soil Erosion: A Field Guide for an Erosion Risk Assessment for Farmers and Consultants*. MAFF Publications, London.

♦ Ensure drainage is well maintained so that excess water can drain away quickly.
♦ Do not compact seedbeds using rollers or over-inflated tyres or high axle loads.
♦ Consider using shallow cultivation techniques so that soil organic matter is concentrated in the surface horizons.
♦ Ensure that additions of SOM are maintained.
♦ Break long slopes with ditches and hedges.
♦ Use grass contour strips to break up the movement of overland flow down-slope, preventing the formation of rills. These strips should be 5–15 m wide and at intervals of 50–150 m depending on the steepness of the slope.
♦ Buffer strips of grass, usually 20 m wide, can be planted along the sides of watercourses to prevent the transport of silt.
♦ Avoid ploughing up-slope and down-slope. Plough lines should, where possible, follow contours and soil should be thrown up-slope. However, if the contours are not followed correctly contour ploughing can make erosion problems worse. If in doubt it is safer to use grass strips.
♦ In areas with high-impact rainfall terracing can be used to reduce slope angle and aid infiltration rates. Figure 7.20 shows an example of small-scale terracing on footpaths in the Peak District, England.
♦ Embankments and blind ditches can be used as emergency measures.

Essential points

♦ Soil erosion is a natural process. Problems occur when the rate of erosion is increased above the rate of soil formation, usually as a result of poor management.
♦ Soil erosion reduces the soil's ability to supply nutrients and water and to provide adequate rooting depth. It also transforms soil from a valuable non-renewable resource into a potential contaminant.

Fig. 7.20 The use of terracing to combat footpath erosion in the Peak District, England. The same method can be used on a much larger scale to prevent erosion at the farm scale.

♦ Erosion is a two-stage process involving the detachment of particles and then their transport to new locations.

♦ Erosion risk can be assessed using a combination of the erosivity of the area and the erodability of the soil.

♦ Soil erosion can be prevented by reducing the time the soil surface is exposed to rainfall, maintaining stable soil structures through additions of organic matter and reducing the potential of overland flow through the use of grass strips, ditches and terracing to break up slopes.

3   How can we measure the 'quality' of the soil?

Measuring 'soil quality'

Recently some scientists have started to develop methods of measuring the quality the soil. The advocates of soil quality measurements sometimes use the phrase 'soil health'. Just as the symptoms of illness, such as heart disease, can be detected by noting blood pressure, cholesterol concentration, obesity and the like, so the belief is that the health of the soil can be assessed using a range of analytical techniques. Soil quality assessments are not a new idea. The US Land Capability Classification system has been copied by many countries as a method of ranking land according to its agricultural quality. Now the aim is to widen the concept of soil quality to include uses other than purely agricultural potential. For example, a conservation agency may wish to assess a soil's conservation or ecological value. In most cases, if we simply were to use the factors which are important to agriculture many upland soils would be regarded as being of poor quality. This completely ignores the fact that they may be valuable for other reasons, such as providing habitat for unique communities of plants and animals.

Although most scientists agree that a measurement of soil quality is a good idea, reaching a consensus on how the measurements should be taken and interpreted has proved less straightforward. This is because the term 'soil quality' can mean different things to different people. As a way of reconciling these interests the current view is that soils should be graded on the basis of 'fitness for purpose'. This means that the quality of the soil will depend on the purpose the soil has been assigned to fulfil.

It is unlikely that there could ever be a single measurement of soil quality. Any quality assessment scheme therefore needs to take into account a range of soil properties. These are sometimes referred to as the 'minimum data set'. A minimum data set might consist of slope, texture, drainage, erosion risk or any number of factors. These measurements are then combined to produce a composite value that consists of a single grade (in the way that a school exam grade is often made up from the marks awarded to several questions).

In order to develop an assessment of soil quality we have to:

♦ Define what is meant by 'soil quality'. In other words, what role are we expecting the soil to perform? This may range from agricultural productivity to upland habitat.

♦ Identify which soil functions are essential in order for the soil to fulfil this purpose. For example, agricultural productivity may depend on soil functions such as nutrient supply, water storage and structural stability.

♦ Identify which soil properties make up each function. For example, nutrient supply is determined by several soil properties such as the concentration of organic matter and soil texture.

♦ Identify methods that can be used to measure each soil property. For example, nutrient supply depends on a combination of CEC and pH.

♦ Develop a grading scheme that combines all the measurements in the minimum data set into a single assessment grade that is readily understood.

Table 7.10 shows how this might be done for assessing the agricultural quality of a soil.

Essential points

♦ Currently scientists are working on methods to measure soil quality.

♦ Soil quality can be defined as 'fitness for purpose'.

♦ It is unlikely that one single measurement will be able to define this. In many countries the minimum data requirements have yet to be decided.

Table 7.10 A hypothetical example of how to assess the agricultural quality of a soil.

Function	Soil functions	Soil properties	Lab test
Arable crop production	Surface factors	Low erosion potential	Erosion assessment (erodability/erosivity)
	Physical factors	Soil porosity Soil texture Aggregate stability Plant-available water	Bulk density Mechanical analysis Wet sieving Moisture release curve
	Chemical factors	pH C and N P supply Base status Pollution free	Titration Loss on ignition $CaCl_2$ extractable P Cation exchange capacity Tests depend on site history

Final word . . .

At the beginning of this chapter we said that we may be about to enter a new biotech age, which in many ways is a technological leap forward that rivals man's first attempts to use tools. The last century saw great advances in the chemical and physical sciences. It is predicted that this century will see similar advances in the biological sciences. The stakes are very high. We now have the ability to engineer new life forms, which have made the whole basis of the 'species' redundant. Ever since the first primitive forms of DNA managed to replicate, the genetic legacy of life has undergone constant change, but these changes have been spread over countless generations. It now faces the prospect of explosive change with the production of 'designer organisms'. This new technology is likely to bring great benefits. There are already developments under way to produce bacteria that can recover metals from low-grade ores and new soil organisms to break down pollutants. However, if history can tell us anything, it is that technological advances always come with a price tag attached. In the next few decades we may be facing the prospect of soil 'genetic pollution' with the development of new, environmentally persistent life forms, which, unlike chemical pollutants, can replicate. This is a problem we have yet to deal with.

Chapter Summary

Soils are a valuable non-renewable resource. In the past soil degradation was often rectified by moving to new, uncultivated land. For much of the world's population this is no longer an option. Soil can be contaminated from a wide range of chemicals. These can be divided into inorganic and organic contaminants. Inorganic contaminants include metals, acid rain and radiation. Organic contaminants include pesticides, PHA and chlorinated compounds. The reactions of organic chemicals in soils are determined primarily by their molecular weight and polarity. A contaminant becomes a pollutant when it poses a health risk or if the biological function of the soil is damaged. Various government bodies publish concentrations of contaminants that constitute a pollution risk. The concentration at which the contaminant is regarded as pollution varies depending on the country. This is because countries vary in their attitude to pollution. We can assess the concentration of pollutants by conducting a risk assessment. In some cases remediation at the site will be possible. This may take the form of land farming or immobilizing the contaminant in plant tissue. Where the pollution is severe, there may be little option but to remove the soil completely. By managing soil contamination we can prevent pollution. In the case of landfill sites, the 'dry tomb' or bioreactor approaches can be used. In the case of sewage sludge, the metal concentration of the soil and sludge can be monitored to ensure that they stay below toxic concentrations.

Soil erosion is a serious problem in many parts of the world. Erosion is a two-stage process, which entails dislodging soil particles, either through water or wind, then moving them to new locations. Soil erosion not only results in the loss of a valuable resource, but also constitutes a pollution risk to watercourses. We can assess erosion risk by noting soil erodability and the erosivity of the rainfall. Erosion potential can be modified by good management practices. These seek to reduce the time the soil is bare, to promote good structural stability and to break up slopes. These soil conservation measures can help maintain the quality and quantity of soils that are vulnerable to erosion.

The development of methods that can assess soil quality has presented a number of problems – this is mainly because 'soil quality' is highly dependent upon what role are we expecting the soil to play. The concept of 'fitness for purpose' allows soils to be graded for a variety of different uses. The quality of the soil is unlikely to be described adequately with a single measurement. Before measurements are taken, the nature of the minimum data set needs to be decided. Measurements are then combined to produce a grade, which relates to how well the soil meets our requirements.

Further Reading

We realize that some of the books listed below may be out of print; however, many can still be obtained through inter-library loan so should still be accessible.

General soil science textbooks

Brady, N. C. (1990) *The Nature and Properties of Soils*. Macmillan, New York. Still the best general soil textbook, comprehensive and written in a clear style.

Courtney, F. M. and Trudgill, S. T. (1993) *The Soil: An Introduction to Soil Study*. Hodder & Stoughton, London. A very good introduction to soil science for students with no scientific background.

White, R. E. (1997) *Principles and Practice of Soil Science: The Soil as a Natural Resource*. Blackwell Science, Oxford. Provides good coverage of all the main subject areas (assumes some scientific background).

Wild, A. (1993) *Soils and the Environment: An Introduction*. Cambridge University Press, Cambridge. Provides a good general coverage with a useful section on how soils interact with the wider environment (assumes some scientific background).

More comprehensive general soil science textbooks

Sumner, M. E. (2000) *Handbook of Soil Science*. CRC Press, Boca Raton, FL. A very good publication for final year and research students. The text is up to date and covers both the agricultural and the environmental aspects of soil science.

Wild, A. (1988) *Russell's Soil Conditions and Plant Growth*. Longman Scientific and Technical, Harlow. Good general text first published in 1912 by E. J. Russell. Since then it has undergone many revisions, so that it still provides a comprehensive source of information focused on soil science from an agricultural perspective.

Web-based general introductions to soil science

Introduction to Soil Science and Soil Resources
http://www.pedosphere.com/

University of Alberta
http://www.environment.Ualberta.Ca/SoilsERM/resource.html

University of Valdosta
http://www.valdosta.edu/~grissino/geol3710/

University of Cranfield Basic Soil Science (for schools and colleges)
http://www.silsoe.cranfield.ac.uk/nsri/

University of Adelaide Basic Soil Science (for schools and colleges)
http://www.waite.adelaide.edu.au/school/Soil/index.html

Soil Science Glossary
http://www.soils.org/sssagloss/

Secrets Hidden in the Soil
http://Ltpwww.gsfc.nasa.gov/globe/

For web searches on soil information

Soil Jumpstation
http://homepages.which.net/~fred.moor/soil/links/l01.htm

International Soil Reference and Information Centre
http://www.isric.nl/

Other more specialized information for each chapter

1   Rocks to Soil

Atlas of minerals
http://www.geolab.unc.edu/Petunia/IgMetAtlas/mainmenu.html

British Geological Survey
http://www.bgs.ac.uk/

US Geological Survey
http://www.usgs.gov/

Photographs of soil profiles
http://soils.usda.gov/

2   Particles, Structures and Water

Hanks, R. J. (1992) *Applied Soil Physics: Soil Water and Temperature Applications.* Springer-Verlag, New York.

Hillel, D. (1982) *Introduction to Soil Physics.* Academic Press, Harcourt Brace Jovanovich, San Diego, CA.

Soil texture and structure
http://syllabus.syr.edu/ESF/RDBRIGGS/FOR345/labtext03.htm

Soil textural triangle
http://www.bsyse.wsu.edu/Saxton/Soilwater/

3   Soil Surfaces, Acidity and Nutrients

Cresser, M., Killham, K. and Edwards, T. (1993) *Soil Chemistry and its Applications.* Cambridge University Press, Cambridge.

Goulding, K. and Annis, B. (1998) *Lime, Liming and the Management of Soil Acidity.* The Fertiliser Society, London.

University of New Hampshire
http://pubpages.unh.edu/~harter/soil702.html

Purdue University
http://www.agry.Purdue.edu/ext/forages/Publications/ay267.htm

Soil acidity and liming (Fred Moor's website)
http://homepages.which.net/~fred.Moor/Soil/ph/p01.htm

4   Soil Microbes and Nutrient Cycling

Killham, K. (1994) *Soil Ecology.* Cambridge University Press, Cambridge.

Lewis, O. A. M. (1986) *Plants and Nitrogen.* Edward Arnold, London.

Richards, B. N. (1987) *The Microbiology of Terrestrial Ecosystems.* Longman Scientific and Technical, Harlow.

Stevenson, F. J. (1986) *Cycles of Soil, Carbon, Nitrogen, Phosphorus, Sulphur, Micronutrients.* John Wiley and Sons, New York.

Wood, M. (1989) *Soil Ecology.* Blackie, Glasgow and London.

Soil food web
http://www.rain.org/~sals/ingham.html

5   Soil Survey, Classification and Evaluation

van Engelen, W. P. (2000) SOTER: The World Soils and Terrain Database. In: *Handbook of Soil Science* (ed. M. E. Sumner). CRC Press, Boca Raton, FL.

Landon, J. R. (1991) *Booker Tropical Soil Manual: A Handbook for Soil Survey and Agricultural Land Evaluation in the Tropics and Subtropics*. Longman Scientific and Technical, Harlow.

McRae, S. G. (1988) *Practical Pedology: Studying Soils in the Field*. Ellis Horwood, Chichester.

Natural Resources Conservation Service (US)
http://soils.usda.gov/

Soil Survey (UK)
http://www.silsoe.cranfield.ac.uk/sslrc

6 Soils and Agriculture

Davies, B., Eagle, D. and Finney, B. (1993) *Soil Management*. Farming Press, Ipswich.

Tillage options
http://web.aces.Uiuc-edu/faq/faq.pdl?project.id=18

Compaction and drainage
http://www.ag.ohio-state.edu/~ohioline/b301/index.html

Michigan State University – Soil and soil management
http://www.msue.msu.edu/msue/imp/modf1/masterf1.html

7 Soil Contamination and Erosion

Department of the Environment, Central Directorate on Environmental Protection and Royal Commission on Environmental Pollution (1985) *Controlling Pollution; Principles and Prospects*. HMSO, London.

MAFF (1999) *Controlling Soil Erosion: A Field Guide for an Erosion Risk Assessment for Farmers and Consultants*. HMSO, London.

National Soil Erosion Research Laboratory
http://topsoil.nserl.purdue.edu/nserlweb

Erosion control
http://www.forester.net/ec.html

Soil health
http://www.ibiblio.org/farming-connection/soilhlth/home.htm

Other useful links

Soil Science Society of America
http://www.soils.org/

Australian Society of Soil Science
http://groundwater.ncgm.uts.edu.au/asssi/

British Society of Soil Science
http://www.soils.org.UK/

Canadian Society of Soil Science
http://www.csss.ca/

New Zealand Society of Soil Science
http://nzsss.rsnz.org/

US Department of Agriculture
http://www.usda.gov/

US Environmental Protection Agency
http://www.epa.gov/

United Nations Food and Agriculture Organization
http://www.fao.org/

Rothamsted Experimental Station
http://www.rothamsted.bbsrc.ac.uk/iacr/tiacrhome.html

National Society of Consulting Soil Scientists, Inc:
http://www.nscss.org/

Index